D0558754

VOLUME 15 IN THE SERIES
Our Sustainable Future

SERIES EDITORS
Charles A. Francis, *University of Nebraska–Lincoln*
Cornelia Flora, *Iowa State University*
Paul A. Olson, *University of Nebraska–Lincoln*

CAROLYN JOHNSEN

Raising a Stink

The Struggle over Factory
Hog Farms in Nebraska

University of Nebraska Press
Lincoln and London

© 2003 by the University of
Nebraska Press. All rights reserved. Manufactured in the
United States of America. ∞
Library of Congress. Catalog-
ing in-Publication Data.
Johnsen, Carolyn, 1944—
Raising a stink: the struggle
over factory hog farms in Ne-
braska/Carolyn Johnsen.   p.
cm.—(Our sustainable fu-
ture; v. 15) Includes biblio-
graphical references and index (p.). ISBN 0-8032-7617-6
(pbk. : alk paper) 1. Swine—
Nebraska 2. Swine—Eco-
nomic aspects—Nebraska. 3.
Pork industry and trade—
Nebraska. I. Title. II. Series.
SF395.8.N2J64 2003
636.4'009782–dc21
2002043024

Contents

List of Illustrations   vi

Acknowledgments   vii

Introduction   ix

1   If Economics Rule   1

2   Hog-Wild to Expand   13

3   Riled Up   18

4   Home Rule   29

5   The Legislature Weighs In   44

6   Hog Hiltons and Initiative 300   56

7   A Tale of Two Counties   68

8   The Marshal Comes to Dodge   83

9   Pork Tenderloin at the Capitol   95

10   Another Pass at the Legislature   105

11   Building on Sand   112

12   The Smell of Money   125

13   To Make a Silk Purse out of a Sow's Ear   135

Notes   149

Index   173

Illustrations

Following page 80

Ranchers Jim Lawler, Ron Lage, and Barb Rinehart

Max and Willard Waldo at their farm near DeWitt

Mabel Bernard enjoys her flowers

Wayne Kaup adopts new methods in Holt County

Brian Mogenson at one of his hog farms in Antelope County

Aaron Spenner and friend

Bob Spenner in his hog pasture near West Point

Finisher pigs at the University of Nebraska hog farm near Mead

One National Farms site in Holt County

Ron Schooley joins other factory-farm opponents

A sow and her litter at Progressive Swine Technologies

Hog farms become a focus of Nebraska news

"If You Can't Beat 'Em, Move"

Elaine Thoendel and Donna Ziems

Acknowledgments

Many thanks to Nancy Finken, news director at the Nebraska Public Radio Network, who permitted me to take a leave of absence to write this book; to Paul Olson, Chuck Francis, and Clark Whitehorn for their enthusiastic support and advice; and to Bud Pagel, Bill Kloefkorn, and Sandra George, who made helpful suggestions on early drafts of the manuscript. Thank you to the Fund for Investigative Journalism for the generous grant that supported my research and to the Wesleyan Writers who offered both comic relief and monthly comments on my writing. My deepest appreciation goes to my husband, Dave Fowler, whose love, advice, and encouragement have sustained this effort.

The essential project of the American West was to exploit the available resources.
– Patricia Nelson Limerick, *The Legacy of Conquest*

Introduction

What I learned of the contemporary swine industry began in 1997 at Nebraska Public Radio. On a beat that encompassed both agriculture and the environment, I naturally paid attention to the growing storm in the countryside. The *Norfolk Daily News* called it "a vitally important struggle . . . for the soul of northeast Nebraska's towns and the farms that undergird them."[1]

My first personal encounter with the intensity of the debate came in October 1997 at a long hearing in a public hall in Crete that was part of a study by two committees of the Nebraska legislature. Among the two hundred or more people in the crowd, the lines were clearly drawn between those who saw opportunity in factorylike hog production and those who predicted the death of the family farm and the contamination of precious air and water. The same actors are still on the stage today, more than four years later.

Between 1997 and 2000 I wrote more than one hundred stories on the controversy over mega–hog farms. It became clear to me that there was much more to the story than was possible to tell in ten or a hundred five-minute radio reports. This book is the result of my desire to know more than I could learn under the pressure of a daily deadline. I also wanted to create a permanent record of a story that turns out not to be about an environmental catastrophe but about a struggle pitting property rights against the right to clean air and water, about rural people's desire to protect their way of life from the impact of big pork producers seizing an opportunity for profit, and about policy makers coming to terms with the need to regulate one of the state's most important economic sectors.

I had read the *Raleigh News and Observer*'s series "Boss Hog" and the continuing coverage after that series, so I knew something of the issues beyond Nebraska. North Carolina's story was dramatic—the hog baron who manipulated the legislature into passing laws protective of his industry, the breached lagoons with manure fouling the waters, the fish kills, and then in 1999 the horrific pictures of hog carcasses floating in rivers after Hurricane Floyd. Some Nebraskans feared that such a catastrophe could happen here.

We don't have hurricanes, but Nebraska does have numerous waste lagoons built in drainageways with millions of gallons of waste held back by dams. Although Nebraska's story wasn't dramatic in the same gut-wrenching way as North Carolina's, there was plenty of drama as Nebraskans who objected to the trend toward concentrated livestock production confronted others who believed it was inevitable and were eager to cash in. The struggle left public officials from county zoning commissions to the governor's office trying to determine an appropriate and fair response.

Aware of the intense debate over what constituted good science, I read peer-reviewed studies and used them as primary sources. Although there is still a need for in-depth research into the effects of concentrated swine production on the environment and human health, many such studies are available to help inform debate on the issues. I've tried to present the scientific information in a way that will help readers understand how citizens can legitimately arrive at differing opinions on the effects of concentrated swine production.

I wanted to do more with this story than what has been described as "referee journalism," where reporters merely give equal space to the "sides" of an issue—a "he said, she said" approach to reporting.[2] Instead, I was interested in using narrative techniques to trace the history and progress of the story, including the incremental decisions that led to where Nebraska is today. In addition to interviewing public officials, I visited with rural Nebraskans in their homes and in coffee shops, courthouses, cornfields, and hog sheds, seeking to put a face on the story. Most people were eager to talk, but four hog producers whose operations were among those at the center of controversy either rejected my requests for interviews or declined to return numerous phone calls requesting interviews. For the sake of balance and fairness, I've drawn their points of view as much as possible from the public record and from interviews with people sympathetic to their operations. I was dismayed and frankly surprised by former governor Ben Nelson's decision not to be interviewed for this book. After all, most of the very large hog operations came to Nebraska during his term in office.

Observing zoning commissions and county boards struggle with the complexity of the issues, I've gained a new respect for local government, whose elected officials live daily with the results of their votes. By comparison, state senators are often out of touch with the local consequences of their decisions. (It must be pointed out that the legislature banned smoking in the capitol to protect public health but did nothing to protect the neighbors of big hog farms from odor that made them ill.)

There are a number of things this book doesn't do. Developments in dairy, poultry, and beef production parallel those in the swine industry. But because mega–hog farms created the controversy in Nebraska and because I had less than a year to write the book, I chose to focus only on hogs for this project.

This book does nothing to explain federal farm policy or the farm program. Also, I've purposely avoided the debate over animal welfare because it would have detracted from the more important issues facing Nebraska. The daily press is doing a thorough job of covering the growing concern over the effects of antibiotic use in confined livestock, so I've only touched on that topic here.

The story of factory hog farms in Nebraska echoes long-running themes in the history of farming on the northern Great Plains. From the beginning, small farms have been replaced by bigger farms. The first farmers in Nebraska were the Pawnees, who raised corn in the valleys of the Platte and Loup Rivers on plots smaller than one acre. The original Homestead Act enabled European settlers to acquire 160 acres. Moses Kinkaid, a U.S. senator from O'Neill, soon persuaded Congress to increase the allotment to 640 acres to accommodate ranchers who couldn't make a living by running cattle on 160 acres.

Pigs and corn were connected long before the 1990s. J. Sterling Morton, founder of Arbor Day and president of the State Board of Agriculture, was "an avid hog-corn enthusiast" who encouraged farmers with the memorable aphorism "Corn is King, Swine Heir Apparent."[3]

Historian Patricia Nelson Limerick has described western history as "an ongoing competition for legitimacy—for the right to claim for oneself and sometimes for one's group the status of legitimate beneficiary of Western resources."[4] This theme manifests itself in the dispute over who is the "real" farmer: the small, diversified farmer who struggles from harvest to harvest with spouse, daughters, and sons working side by side on the land and caring for the livestock, or the big multicounty or multistate operator who hires others to do the labor and spends his time managing the books and lobbying politicians to arrange easier ways to make a profit.

In 1998, as the controversy over big hog farms was just picking up in Nebraska, a *New York Times* editorial framed the issues in this way: "The battle over hog factories is really a battle over the definition of farming and the legal consequences that follow from that definition."[5]

As this book goes to press, that battle continues nationwide; the legal consequences will be years in the making. The controversy is in flux, with new information almost daily. *Raising a Stink* will lack some of that information, but the book wasn't intended to be as timely as today's news. Instead, it is meant to provide a lasting record of the turmoil that engulfed many small communities in Nebraska at the turn of the millennium. Finally, I hope it encourages readers to consider the implications of where and how hogs are raised for the bacon, sausage, ham, and pork barbecue we eat.

Farm families represent the best of America. They represent the values that have made this country unique and different—values of love of family, values of respect for nature. I always tell people that every day is Earth Day when you own your farm, when you're working the land.

– President George W. Bush, remarks upon signing the Agriculture Supplemental Bill at the Bush Ranch, Crawford, Texas, 13 August 2001

If Economics Rule

MABEL BERNARD

To drive through Chase County, Nebraska, in June and July is to marvel at how, in every direction, irrigation pivots spread water like long rows of gossamer curtains over the green corn. With an annual production of more than twenty-two million bushels of corn and an abundant aquifer, Chase County is also a likely place for pigs.[1] Confinement hog production—the newest industry in the county—is driven by a tight loop of interdependence involving hogs, water, and corn. Hogs eat corn; their manure and urine are mingled with water and shot through center pivots onto corn to produce feed for more hogs. In 1997 there were only about fifty-six hundred pigs in the county; by 2001 that number had risen to more than eighty thousand.[2] The Imperial Chamber of Commerce and some county officials welcomed the prospects for new business and new property-tax revenue.

Meanwhile, ninety-four-year-old Mabel Bernard was left to cope with the effects of thirty-six thousand hogs that moved in near her home south of Enders. Eighteen long white hog barns sit amid cornfields about a mile north of the small white stucco house where Mabel lives. A thin grove of trees stands on the north side of her house—a common sight in the Great Plains, where the wind blows on most days. In 1993 a tornado uprooted many trees, but those that remain may intercept some of the stench that rolls down from the hog farm when the wind blows from the north.

1

Mabel lives alone in this house just inside the Dundy County line, but she often has visitors. Her daughter stops in two or three times a week on her way to carry water to cattle in a nearby pasture. Mabel's neighbor and distant relative Joyce Bernard often visits her. On a warm spring day in 2001, Joyce joins Mabel in her living room and talks about what it is like to live here. The air is blessedly free of the odor of pigs.

Dressed in a loose blue smock and low soft shoes, Mabel is seated in a rocker, her short white hair like a halo around her face. She broke her leg eight years ago and still gets around with the help of a walker, which stands next to her chair. A pile of knitting and newspapers lies on the floor, and the telephone is within reach on a low table. Mabel maintains a sweet half-smile as she talks softly about her life before and since the hogs moved in.

"This is my in-law's homestead," says Mabel. "They came here in a covered wagon in 1886—over a hundred years ago."[3] In 1926 Mabel and her husband, George, moved here. They built this house in 1948, replacing the sod house that George's family lived in. The couple raised three children on this homestead. Mabel's father-in-law was bedfast for forty-one years; she helped care for him twenty-three of those years until he died in 1948. George died in 1973 and Mabel has since lived here alone.

"I think I'm entitled to clean air," she says.

Tim and Steve Leibbrandt—whom Mabel had known and liked since they were children—stopped in one day in 1998 to tell her they were planning to operate a hog farm on the hilltop to the north. Mabel's cousin sold the land to the Leibbrandts. "Then he moved to California," she says. "I don't know if he knew what would be here." The Leibbrandts told her the pigs wouldn't smell. In the typically generous manner of many rural Nebraskans, Joyce and Mabel both say that perhaps the Leibbrandts didn't know how bad the odor would be. Steve Leibbrandt lives about three miles from the operation, and his brother Tim lives about twenty miles away, north of Imperial.

Sometimes the pigs smell for three or four days in a row and then not for several weeks—usually depending on the direction of the wind. The odor invades Mabel's life often enough to have generated strong feelings in this gentle, frail woman. "I feel cheated because I've been here so long," says Mabel. "I think I deserve more consideration."

But for her game leg, Mabel says, "I've been blessed with good health." The pigs have changed that. With a shifting wind, the stench of thousands of hogs sometimes wakes her up at night, burning her eyes and making her sick to her stomach.

Joyce says, "It's the sewer we smell."[4] She uses the word "sewer" in the same spirit that neighbors of other such operations have angrily called the big open pits where hog waste is stored "cesspits." But neither Mabel nor Joyce has complained to the Leibbrandts. "Everyone is polite and no one complains," says Joyce. "There's nothing to do. They've invested all this money, so they're not gonna stop."

"I burn candles and that helps," says Mabel. "Some." She can't drive so has no option but to endure the stench until the wind shifts. Even then, the smell of pigs clings to fabrics and other surfaces in the house. Mabel doesn't want to move. "This is my home. I have no other home," she says. "I love my home and it means a lot to me."

With a persistent, gentle smile, Mabel Bernard talks about her family's life on this farm that has never been mortgaged. She recalls her husband telling of the time his mother hid the children under the kitchen table to protect them when stampeding buffalo crushed the corner of their soddy. Now her grandson rents some of Mabel's pasture for his cattle—the fourth generation on this Dundy County land.

The conversation turns to Mabel's garden and she offers to show it off. She dons a wide-brimmed straw hat with a jaunty pink ribbon. Getting a grip on her walker, she pulls herself up from her chair and walks to the kitchen, where she sets the walker aside and grasps a three-legged cane left there to help her down one step and through the back door. Once outdoors she trades the cane for another walker to move among the hollyhocks, lilies, and larkspur that bloom on this fine June morning. Mabel spends less time in the garden now than she once did because the odor from the hog farm often makes it unbearable to be outdoors. Mabel has always had a big flower garden, but she says, "I can't plant enough flowers to drown it out." She beams at her garden and at the field of ripening wheat beyond and says, "We'll be harvesting by the Fourth of July."

Mabel would like to linger in the garden but agrees to let Joyce help her out of the heat and back inside. Although there's no smell of pigs on this particular day, it's easy to believe Mabel Bernard's stories of how thirty-six thousand pigs have changed her life. She is also concerned about how the waste from so many hogs—stored in vast earthen pits—will affect the groundwater if it seeps in and the surface water if it runs over.

If she ever chose to contact her state senator or any agencies of government with the words "health" or "environment" in their titles, Mabel Bernard would be one voice among many that have risen from the Nebraska countryside in recent years to say "I think I deserve more consideration." The Nebraska legislature has chosen not to regulate odor, leaving it up to the counties to limit the effect on citizens by imposing minimum distances between big livestock operations and their neighbors. Chase County has no authority over the Leibbrandts' operation because it was built before the county adopted zoning regulations.

But on the other side of the issue are hundreds of pork producers who see an opportunity for unprecedented profit by raising pigs in confinement. For them, the long white barns dotting the Nebraska countryside represent a natural step in the progress from wasteful subsistence methods of raising pigs to efficient modern methods. Despite numerous requests for an interview, Steve and Tim Leibbrandt and others involved in bringing confinement feeding to southwest Nebraska declined to explain their point of view on the issues. Some other hog producers are more willing to talk about their enthusiasm for raising pigs in confinement.

JIM PILLEN

In the springtime the rolling hills west of Columbus in northeast Nebraska are dark with the promise of moist earth waiting for seed. In green pastures new calves scamper among patient cows, and watchful hawks hunker down on telephone poles. In May 1998 Jim Pillen drove me through this countryside to look at some of his hog operations—places with names like Cedar Rapids Finisher, Beaver Valley Reproductive Center, Mount Echo, and Pheasant Ridge. Pillen was eager to show off the "farms" that had placed him among the fifty biggest hog producers in the United States—seven of whom operate in Nebraska. In 1997, with fifteen thousand sows, Pillen first made *Successful Farming*'s list of the top fifty "Pork Powerhouses." By 1998 he owned 27,500 sows and was growing.[5] Twenty-five hundred of those sows were housed in the P.C. West Reproductive Center, about twenty miles northwest of Columbus.

The windows are closed on the pickup, but as we drive up a curving gravel road to the long white buildings at the top of the hill, I can smell the hogs. When Pillen parks and I open the passenger door, the odor makes me cough, and my eyes water. Seemingly unaffected, Pillen tells me the waste from the operation is managed well, so I won't notice any odor. Still coughing, I contradict him; he grins, shrugs, and says he supposes he's become used to it. Flies land on my arms and hands and face as we walk toward the office. I hear meadowlarks calling to each other in the distance, but no squealing of hogs. The silence puzzles me because there are twenty-five hundred sows in these barns, but Pillen says the pigs have no need to squeal because they are contented.

Anyone entering the barns must first shower. Pillen asks my size, and from a cabinet on the wall pulls a red jumpsuit, clean underwear, socks, and boots. He points me to a shower and dressing room with towels and antiseptic soap and tells me to wash everything—head to toe—and change clothes before I enter the barn. I'll have to shower again on my way out. The routine is part of an elaborate system of "biosecurity" designed to prevent the introduction of disease.

After showering and donning "reds," we pass through a couple of doors, closing them behind us, and into a vast room where our entrance stirs up a cacophony of grunts and squeals. In this airy concrete-and-steel space, hundreds of sows are uniformly housed in pens that Pillen says are twenty-three inches wide and seven feet long—too narrow for the sows to turn around but just long enough for them to move forward or backward a couple of steps and to lie down. Their snouts are always just a few inches from troughs holding feed and water.

Each animal has its own feedbox where it is automatically fed. Pillen says, "We know down to the tenth of a pound how much feed is being fed so that we can make sure of the animal's exact body weight and metabolic rate so it can be reproductively successful."[6]

Pillen, who calls his business "Progressive Swine Technologies," is effusive about the impact of technology on swine husbandry. He tells me that pigs raised in dirt lots or on pasture need twelve bushels of corn and seven months to grow to a market weight of 220 pounds. In contrast, he says he can raise an animal to 280 pounds in five months on seven bushels of corn in confinement.

"We've taken advantage of genetic technology and production plant technology to really start to tap into the genetic and biological potential of the pig," says Pillen.

Genetic manipulation has obscured the characteristic colors of Landrace, Hampshire, and Duroc breeds in a new genetic strain that Pillen calls c22. The sows are all a uniform pinkish color, with only an occasional black spot to reveal the presence of a Hampshire ancestor in the genetic line. Only the individual paper records attached to the frame of each stall indicate any variations among individuals. The National Pork Producers Council touts the uniformity of the modern hog—leanness that appeals to consumer tastes, a predictable rate of gain that allows producers to control costs, and uniform size and quality that give meatpackers an edge in processing and selling.

Each sow is bred by artificial insemination; computer spreadsheets track her progress from estrus to delivery. She has no contact with other sows except through the bars of the pens; low-level doses of antibiotics help guard against disease. There are no changes in the weather, no fighting to get to the feed or water trough (which explains the lack of squealing), no mudholes to wallow in. Confined swine look very clean.

Fans circulate the air; climate sensors keep the building at an even temperature. Beneath the slatted flooring of the pens, concrete pits two feet deep collect manure and urine in water until it is flushed to a holding pit outdoors. In this room I smell a trace of ammonia and assume that the powerful odor outdoors comes from the waste pit.

Pillen says the clean, sheltered environment makes for happy pigs and more profits. "We've learned to handle an animal so that it is not stressed and so that it is in an environment that is friendly to the pig, which allows that animal to grow at a more efficient rate." Efficiency is a byword of the industry—from breeding to packaging.

For birthing—or farrowing—sows are moved to another area where each sow has its own pen and a heated mat on the floor to keep her litter comfortable. Baby pigs are left with their mother to nurse for about three weeks and are then weaned. A few days later the sow is bred again. These animals are like reproductive machines, bred for the first time at eleven months of age and continuing the cycle until, after four years and eight or nine litters, they have outlived their usefulness and are sent to market.

Each year, the sows in these barns will produce more than sixty-two thousand baby pigs, genetically designed to transform corn into lean muscle as quickly as possible. The weaned pigs are moved to a nursery where they are fed for a few weeks to put on a little more weight, and then moved to finishing barns to grow to about 250 pounds—the last stop before slaughter. The entire process—from birth to slaughter—takes about five months.

Even when hog prices dropped to record lows, as they did in 1994 and again in 1998, operators like Jim Pillen continued to profit. Efficiency of scale in confinement operations and contracts with packers guaranteed them a decent price in return for timely delivery of hogs for slaughter.

"Economics rule," says Pillen, pausing near a sow and eight or nine nursing piglets. "This business is real simple. Whatever's good for pigs is good for people. Whatever's good for people is good for pigs. If this wasn't good for this pig, this pig wouldn't be productive and we wouldn't have a very good business."[7]

But critics charge that confinement profits come at the expense of people and the environment. The high-tech processes involved in a big confinement operation don't extend much past the pipe that takes the manure and urine to the waste pit—euphemistically called a "lagoon."

An adult hog produces two to five times the amount of waste that a human being produces.[8] The twenty-five hundred sows at P.C. West produce waste equivalent to that of a city of at least seventy-five hundred people. Unlike human waste, which in cities is treated to remove pathogens and other pollutants, hog waste typically receives no treatment. Instead, solids settle to the bottom of the lagoon where anaerobic bacteria (which thrive in the absence of oxygen) gradually digest them; the liquid on top—which includes the flushing water—is periodically sprayed onto cropground through center pivots. Nitrogen-rich hog waste can be a valuable fertilizer for corn, which in turn feeds the hogs—an efficient, logical cycle that seems especially suited to Nebraska, with its thousands of acres of corn and abundant water and land. But even with careful construction and management, these systems can cause serious problems.

In proposing a national strategy to regulate livestock operations, the U.S. Environmental Protection Agency said that facilities with more than one thousand animal units (twenty-five hundred hogs) "produce quantities of manure that are a risk to water quality and public health whether the facilities are well managed or not."[9] Those risks include runoff when the effluent is sprayed onto crops, emissions like ammonia and hydrogen sulfide that rise from the lagoon's surface, leakage into groundwater, and overflows that contaminate surface water.

All states allow some amount of seepage from hog waste lagoons; until 1999 Nebraska permitted all lagoons to leak a quarter inch per day—one of the highest rates in the nation.[10] For each acre of lagoon surface area, this amounts to nearly sixty-eight hundred gallons of seepage per day.[11] Most of the hundreds of hog waste lagoons in Nebraska are lined with only a one-foot layer of compacted clay, which livestock producers and some scientists say is sealed by manure within a few weeks of use.[12] To limit seepage, Pillen lines his lagoons with a clay liner covered by a forty-millimeter-thick plastic liner, which he says doubles or triples the cost of building the lagoon.

The one lagoon at P.C. West is multiplied by four at Wolbach Foods, where Pillen sends his nursery pigs to grow to market weight. At Wolbach Foods, twenty-four long white barns with silver feed bins at each end create an industrial presence on a section of what once was pasture and cropground. This operation feeds out about 115,000 hogs each year, which produce waste equal to a city of about 400,000 people. The hog waste is flushed from the barns into four big lagoons with a total surface area of about twenty-seven acres.[13]

On the summer day when I visit, the only noticeable odor seems to come from inside the hog barns. The feeder pigs here have more freedom, running in groups of about twenty in larger pens in barns where side curtains are pulled up to let the air through. Considering the scale of these operations, I asked Darin Uhlir, who manages Wolbach Foods and its fourteen employees, if he considers himself to be a farmer.

"Yeah, it's a farm, with an incredible business twist and the need for people skills," he says. "It's everything I ever wanted in a job. I love helping people grow. Our business is developing people and raising pigs. Like my dad. If you ask my dad what was his mission as a farmer, he'd say raising cattle and raising kids."[14]

Uhlir, a tall, muscular man in his late twenties, grew up on his family's ranch near O'Neill and says he always thought he'd make his living on the ranch, but circumstances led him to the hog industry. He says he makes a good living working as a manager for Pillen, who lives about sixty miles away in Columbus.

"I'm one of hundreds of good Nebraska farm youth," says Uhlir. "Half the guys who work out here—if the farm economy had stayed like it was in the '70s and '80s—we would have bought 640 acres and 150 cows, a couple of hundred hogs and grow some corn and hay and make a living. But that's no longer an option. Yet all those die-hard traditionalists want me to go out and buy ground and a big tractor, but the revenue is the same as it was thirty years ago and you need a lawyer to deal with the government regulations. If I wanted to do a hog farm, I wouldn't know where to begin. It's almost intimidating to go into farming. It wouldn't pay to take care of fifty cows and two hundred acres of corn and make thirty thousand [dollars] a year."

Big operations like Pillen's have captured the majority of the hog business in Nebraska and nationwide. Through contracts with big packing plants like IBP in Madison and Farmland in Crete, big producers are guaranteed a market and a price for their hogs—a widespread practice that virtually shuts out smaller operators. In 1970 Nebraska had about thirty-one thousand hog farmers. By the year 2000 the number had dropped to four thousand. Most of those who quit had raised a few hundred hogs per year. The number of Nebraska farmers raising more than two thousand hogs a year increased from 170 to 320 in the last three years of the century. The trend is similar nationwide.

Between 1970 and 2000 more than 90 percent of the hog farmers in the United States quit raising hogs; the numbers of hog farmers in that time fell from 871,200 to 77,860.[15] But in that same time, total hog production rose from 13.4 billion to 19 billion pounds. By 1997 farms selling more than fifty thousand hogs a year—just 1 percent of the total number of hog farms— accounted for 37 percent of total U.S. pork production.[16]

Some small producers persist in spite of fluctuating prices and competition for markets. In Nance County, about fifteen miles east of Wolbach Foods, a sign on Highway 14 north of Fullerton advertises "Wetovick's Swine Farm." The accommodations that Kevin Wetovick provides for his hogs are in a centuries-old tradition of swine husbandry. His purebred Yorkshires range in outdoor dirt pens with sagging, cobbled fences held together with baling wire. Depending on the season, they're given some freedom to graze on grass or grain stubble. Three or four boars are kept in the tightest pen on one side of a weathered shed. On the other side are fifty or sixty feeder pigs. The animals in both pens have access to shelter in the shed and to the out-of-doors. In each pen recent rains have created a mudhole where at least one pig lounges.

Wetovick, who wears a t-shirt that says "Just Gotta Farm," gives me a pair of plastic bags shaped like boots to put over my feet to protect his pigs against any disease I might carry. This is his equivalent of the biosecurity I witnessed at P.C. West. Wetovick tells me that a more elaborate confinement operation would require an investment he isn't able to make.

"I raise the hogs outdoors because I have a very low amount of capital," he says. "The last thing I'm going to do is invest in climate-controlled buildings, other than my farrowing barn."[17]

The farrowing barn is a room in the middle of the shed between the shelters for boars and feeders. Wetovick opens the door—which is latched with a twist of baling wire—and invites me to step inside. On this mild July afternoon, a big fan provides "climate control" for two sows and their litters. They are confined to battered wooden farrowing crates that still look serviceable, although Wetovick tells me they are ten years old. About twenty weaned pigs are penned on the opposite side of the room. The slatted metal floors of the farrowing crates allow waste to drop to the concrete floor a few inches below. There are some flies in this room but very little odor. Wetovick says he hoses the waste out the back of the barn about once a week.

These pigs look healthy and well cared for despite what some would consider primitive accommodations. Yet Wetovick speaks the lingo of the swine industry when he talks about his hogs. He says, "The challenge is to breed a modern hog that is lean and will make the customers money because most all the hogs now are bought by the packer on the basis of lean value and that means you're going to have to have low backfat and a lot of muscle."

He says his hogs don't always fit those criteria. "It's a challenge to breed hogs that can run around and move freely and grow in less than total climate-controlled buildings," he says. "So when mine are out in the elements it takes a little tougher durable hog to survive, and sometimes the extreme lean and muscled hog can't survive in an outdoor facility."

He feeds a pen of bred sows by throwing mixed grain onto the dirt; the sows squeal and push at each other to get to the food.

"This might not be as civilized as when a sow's in an individual crate where she gets her own feed and doesn't have to compete for it," says Wetovick. "Mine kind of have to compete, so I just spread it out more so they don't kill each other."

I ask Wetovick what he would say to criticism that his operation stresses the animals unnecessarily. He answers, "Stress can also be caused by being locked in a small pen with your nose right to the butt of another hog. My pigs get dirty and to me they're quite content."

Wetovick sells most of his weaned pigs at about fifty pounds as feeder pigs to other small local farmers—like Ron and Annette Dubas who live a few miles away.

The Dubases raise corn, beans, alfalfa, a few cows and calves, and feed some cattle and hogs to market weight. Until prices dropped in the mid-'90s, they had their own herd of forty to eighty sows. With no profit in sight, they sold the sows. The forty feeder pigs now in residence at the Dubases' farm are kept outdoors on a two-and-one-half-acre dirt lot with a mudhole in the middle. A couple of hogs wallow in the mud, and all the hogs have dry caked mud on them. There's a makeshift shelter made from half a big steel tank that Ron says once held fertilizer or liquid protein.

"When I get them, I keep them in the barn for a month with straw because it's a little warmer than out in these lots and not as drafty," says Dubas, "but I've been raising hogs my whole life and I sort of know what it takes."[18]

Dubas and Wetovick have no lagoons to store hog waste. Under Nebraska law they're too small to be regulated—unless they're found to be polluting surface water or groundwater. The waste produced by these pigs is trampled back into the soil of the hog pens and occasionally scraped off and scattered on fields as fertilizer. Dubas feeds his hogs corn and oats in a self-feeder, which bangs as hogs push their snouts under the hinged lids to reach the grain. Dubas says he doesn't know how much grain it takes to add a pound to a feeder pig.

"I'm not real concerned about the rate of gain," he says. "It's the finished product that I want. It'll take me seven to nine months to get a three-hundred-pound hog."

In recent years it's been hard for Dubas to sell his hogs to packers because he can't guarantee the number or uniformity that packers expect. So for the last year the Dubases have sold about one custom-butchered hog a week in packages of pork chops, sausage, and roast directly to consumers at farmers' markets. In 1999, with other farm families in the neighborhood, they formed a cooperative called North Star Neighbors to sell their pork, beef, and chicken directly to consumers and thus keep more profit for themselves.

Hogs were once considered a dependable source of cash for farm families. Running on dirt or pastures or scouring corn stubble for grain, they required little capital investment, but there was always a market for hogs—often at the local sale barn where buyers would bid on animals for the packers. As "mortgage busters" a few hogs could always be sold to cover debts when grain and other livestock prices fell. "They were the chores and the grocery money and living expense money," says Ron Dubas. "I don't know if anyone calls it that any more, things have changed so much, but it was good for us."

For the Dubases and Wetovick, hogs, like the rest of the farming operation, are a family enterprise—something the Dubases believe is threatened by what Annette calls "mega-farms." She fears that the concentration of so many animals in limited areas will result in serious environmental contamination. She also fears the loss of farming as a way of life that has sustained her family. In 1997 those concerns led the Dubases to join with other farm families in the area to form Mid-Nebraska PRIDE (People Responding in Defense of Our Environment). The group hoped to influence policy makers to slow the confinement trend until the effects on rural life could be better understood.

"We aren't against hogs and we aren't against hog production or hog farmers," says Annette. "Our concerns come from the highly concentrated production of hogs."[19] She points out that many small hog operations scattered across the rural landscape offer little threat to the environment while they preserve a livelihood for farm families who have roots in the community and provide the labor and investment for operations they own.

But for others, the concentrated production of hogs makes perfect sense in Nebraska—with its abundant land, water, and corn. In October 1997 Dorchester hog farmer Syd Burkey urged a committee of the Nebraska legislature not to hinder growth with government regulation. He said, "We must keep developing livestock production in Nebraska to efficiently utilize our vast agricultural resources. If livestock production in our state is devastated by a moratorium or severe restrictions on responsible, cost-effective production . . . the economic benefits of livestock will migrate to other states and countries."[20]

At that same hearing, University of Nebraska–Lincoln animal science professor Rick Koelsch said Nebraska could handle the load of waste produced by big hog operations. "In Nebraska, the environmental problems are much easier to deal with than for a swine producer in North Carolina or a dairyman in New York because they are shipping into their state on a continuous basis nutrients from the Midwest in the form of animal feeds," said Koelsch. "And they don't have the land base available for recycling those nutrients."[21]

For defenders of industrialized hog production, the recycling of "nutrients"—heavy loads of nitrogen and phosphorous in hog manure—accomplished by pumping liquid hog waste through center pivots to irrigate cropground, completes a logical nutrient cycle, from corn to hog to manure and back to corn. It's a concept that in practice means considerable profits for some, and it has formidable champions. For example, Mark Drabenstott, vice president and economist at the Federal Reserve Bank of Kansas City, writes: "Pork operations in the Great Plains have strong advantages. They are near abundant corn supplies in the western Corn Belt. What is more, the new farm bill makes it easier for farmers in the Great Plains to switch from wheat production to crops that are better feedstuffs. . . . Finally, pork operations in the Great Plains can be located in areas with some of the lowest density of population in the nation. Further, many communities in the region offer ecosystems with substantial capacity to handle animal waste."[22]

At the turn of the millennium, in Nebraska and other Great Plains states, Drabenstott's assumptions about cooperative people and resilient ecosystems would be at the center of heated debate. The population was indeed sparse in most counties where the biggest hog farms tried to locate. But many residents refused to accept the premise that their land, water, and air had "substantial capacity to handle animal waste."

We must not lose sight of the imperative to protect our environment as we go about the business of strengthening our economic base. Short-term economic gain without environmental foresight will cost us dearly in the long run.

– E. Benjamin Nelson, First Inaugural Address, 10 January 1991

Hog-Wild to Expand

In the spring of 1997, 44 percent of the hog farms in Nebraska resembled those run by Ron Dubas and Kevin Wetovick, in that they had fewer than two hundred hogs. But a trend toward bigger operations had been evident for more than thirty years. Between 1965 and 1997 more than thirty thousand Nebraska farms stopped raising hogs, while the number of hogs raised in the state rose from about 2.8 million in 1965 to 3.5 million in 1997.[1]

Some of the nation's biggest pork producers contributed to these numbers. In Adams County, Hastings Pork housed about 50,000 hogs in buildings that sprawled across 131 acres of land on a former World War II ammunition depot. Near Atkinson in Holt County, National Farms had two sites with a total capacity of about 150,000 hogs. In three decades of raising pork in Nebraska, companies with connections to Sand Livestock Systems had about forty operations in eleven counties—each of them housing from 2,000 to more than 18,000 hogs.

Over the years, objections to the presence of so many hogs had remained localized. For example, in the 1980s and early 1990s neighbors filed six nuisance lawsuits against National Farms to force the company to reduce odor emitted by its hog farms. The Nebraska Supreme Court's decisions against National Farms in that local dispute would take on wider significance during the statewide controversy that later arose over hog production.

What some would call a "revolution" began with a sharp rise in the number of applications for construction permits received by the Nebraska Department of Environmental Quality (NDEQ) in the spring and summer of 1997. Typically the NDEQ received five to ten applications a month from people wanting permits to build facilities to store manure. In June, July, and August of 1997 hog producers filed eighty-eight permit applications with the NDEQ.[2] More than two hundred construction permits would be requested that year—about twice the usual annual number.[3] Most of the permits would be for operations with more than a thousand hogs. By summer, bulldozers had roared into at least ten counties, digging waste pits that would underlie confinement barns and excavating deep lagoons to hold millions of gallons of hog manure, urine, and water.

Crews working for Premium Farms, a company based in Neligh, began moving earth for buildings to house fourteen thousand feeder pigs near Brunswick in Antelope County. With a thirty-million-dollar loan from a Delaware company, Premium Farms' owner Brian Mogenson announced plans to build at least nine facilities in northeast Nebraska to house fourteen thousand hogs each.[4] In Harlan County, Sand Livestock Systems prepared to build about thirty long barns to feed, house, and handle the waste of thirty-four thousand feeder pigs on a hill northwest of Alma. Bell Farms of Wahpeton, North Dakota, approached landowners in a sixty-square-mile area of Nance and Greeley Counties to buy land to accommodate up to half a million hogs.[5]

In 1997 Nebraska was seventh among the states in hog production.[6] The corn, water, and land base that industrial hog farms needed had always been available in Nebraska. But now other factors would encourage the swine industry to expand in Nebraska.

States with a bigger concentration of hogs had recently received a lot of negative publicity. In February 1995 the *Raleigh News and Observer* ran a series of stories investigating North Carolina's swine industry. The stories revealed that the boom in hog confinements that had made North Carolina second in the nation in pork production had also seriously polluted air and water and overwhelmed the ability of the state's environmental agency to provide oversight. The stories also showed how Wendell Murphy—the "Boss Hog" of the series title and a former state legislator—had manipulated public policy and law to favor factory hog farms.[7] The series, written by a team of investigative reporters, won a Pulitzer Prize for public service. The North Carolina General Assembly had barely begun to respond to the stories by revising the environmental laws for big hog farms when a rainy summer brought more national attention to the dangers of concentrated animal feeding.

In June 1995 an eight-acre waste lagoon at ten-thousand-hog Oceanview Farm broke, releasing twenty-five million gallons of liquefied manure into the New River and killing three thousand fish.[8] Numerous other spills and deliberate discharges of manure followed in the wake of the New River catastrophe. A U.S. Senate report concluded that thirty-five million gallons of spilled animal waste killed ten million fish in North Carolina in 1995.[9]

Prodded into action by the *Raleigh News and Observer* series, the manure spills, and the resulting public pressure for reform, the North Carolina General Assembly passed laws in 1995, 1996, and 1997 imposing more stringent regulations on livestock operations, including permitting, inspections, and odor management. The laws also restricted where operations with more than 250 hogs could be sited and required that neighbors be notified when a hog farm was proposed. In August 1997 North Carolina imposed an eighteen-month moratorium on new or expanded hog farms with more than 250 animals. The general assembly also ordered the North Carolina Department of Agriculture to phase out the use of earthen anaerobic lagoons and sprayfields as the primary way of disposing of animal waste.[10]

Policy makers in other states were also responding to public pressure to regulate factory hog farms. In 1996 "more than 40 animal waste spills (had) killed 670,000 fish in Iowa, Minnesota and Missouri."[11] By 1997 state legislatures in Mississippi, Illinois, Virginia, Kansas, Missouri, Oklahoma, South Carolina, and Minnesota had considered or passed tougher laws governing hog farms. In Iowa, Michigan, Utah, Colorado, Kentucky, South Dakota, Indiana, Georgia, Mississippi, and Texas, activists and some governors called for more stringent laws to protect the water and air from pollution by big hog farms.

Under pressure from environmentalists, the U.S. Department of Agriculture and the Environmental Protection Agency (EPA) began drafting new federal rules for regulating livestock waste.[12] The EPA identified agriculture as the primary source of surface-water pollution in the United States, and zeroed in on the concentrated feeding of livestock as a major concern.[13]

But concerns for the environment had no effect on consumer demand. In 1997 pork dominated more than 40 percent of the worldwide market for meat. That share was expected to grow with increased consumption of meat in developing countries.[14] Hog prices, which had fallen in 1994, were recovering and profits were again possible.

One Nebraska pork farmer, Dan Hodges, said all producers "have been expanding production much faster than the demand was expanding, because we were being told we only had a small window of opportunity to seize the overseas markets that were opening."[15] Hodges said large producers were trying to get new facilities in place before new environmental and zoning laws took effect.

Big hog producers turned to the West for expansion—to Colorado, Idaho, Utah, and Nebraska—where state and local environmental laws were, for the time being at least, more accommodating.

In 1994 and 1995 the Center for Rural Affairs—a small-farm advocacy group in Walthill, Nebraska—had published *Spotlight on Pork*, criticizing corporate hog farms in other states. The authors touted Initiative 300—Nebraska's constitutional ban on corporate farming—as a reason why the state had avoided environmental damage and had lost fewer small hog producers than other states that allowed corporate farming.[16]

But I-300 placed no limits on the size of farming operations. So in 1997, when big hog-producing companies organized as partnerships began building in Nebraska, many people who had read and believed the negative reports about concentrated animal feeding were alarmed. Others merely saw capitalism at work, and opportunities for profits in hogs.

Roy Frederick, a University of Nebraska–Lincoln agricultural economist and former director of the Nebraska Department of Agriculture, said expansion in Nebraska made sense. He pointed out that, during the previous two decades, hogs had been the single most profitable enterprise in Nebraska. "Profitability, especially if it's over an extended period of time, stimulates interest in production," said Frederick. "It's purely simple economics."[17]

In fact, an employee of Sand Livestock Systems—one of Nebraska's most rapidly expanding hog producers—said the company's farms had "produced an average 83 percent return on investment over the years."[18] Developments in genetics, equipment, and nutrition and changes in marketing practices contributed to this kind of profit. Hundreds or thousands of hogs sharing the same genetics, provided with a uniform diet, kept in identical stalls in climate-controlled barns, and sold at a guaranteed price under contract to packers were likely to guarantee a profit to those with enough capital to invest. This is what economists call "economies of scale."

"The larger you got the more efficient the production of each individual animal was," said Frederick, "so that tended to stimulate production in large units as opposed to just a few animals where you wouldn't necessarily get those economies of scale. You put all that together, it's probably not surprising that the industry has gone like it has."[19]

In considering policy, many Nebraska state senators cited Frederick's opinions about the inevitability of the trend toward producing hogs in big confinement units. But studies by other economists have shown that the most efficient hog operations aren't the biggest. They include other factors besides opportunities for profit in judging the effects of feeding thousands of hogs in a limited space.

John Ikerd, an economist at the University of Missouri in Columbia, said industrial methods of producing food threaten the environment, natural resources, and the quality of rural life. Ikerd said, "The same technologies that support our large-scale, specialized system of farming, the industrial systems through which we have increased agricultural productivity, these same technologies have now become the primary focus of growing public concerns"—concerns that Ikerd said were justified.[20]

If the powerful economic forces of industrialization were not to over-whelm social, environmental, and economic values that citizens wanted to preserve, Frederick and Ikerd agreed that policy makers would have to intervene. As in so many other states, the debate over whether and how to intervene would involve every level of government in Nebraska—from township and county boards to the legislature, the governor's office, and the courts.

I know that some people's vision of rural America includes lots of small farms that each have a few acres of land, a few hogs, cows, horses, and chickens running around in the open. Their agenda is to keep the farmers small, and rural America poor and dependent on government handouts. Frankly, these people seem to be on a mission to destroy anyone who can demonstrate that American farmers are capable of being more than peasants or serfs.

– Tim Cumberland, Written Testimony, Legislative Resolution 123, 16 September 1997

Riled Up

The town of Fullerton in east-central Nebraska lies at the confluence of two shallow, slow-flowing prairie rivers, the Cedar and the Loup. Neatly kept houses and gardens along wide, shady streets reflect the efforts of descendants of hardworking German and Polish farmers who settled the community. But peeling paint in several neighborhoods and dark store-fronts on the main street signify a fading glory. In the 1980s and 1990s the town lost a John Deere dealership, a grocery store and meat locker, a clothing store, a hardware store, a coffee shop, and a gas station. But as the Nance County seat, Fullerton, population about fourteen hundred, still had an edge over other towns in the area. In 1997 BZ's Cafe still served three meals on most days and numerous helpings of coffee and pie in between. There were still two locally owned banks that for most of a century had depended on the surrounding family farms for their own prosperity. "They're good to farmers and farmers are good to them," said Lawrence Klassen, who has farmed in the Fullerton area for most of his seventy-two years.[1]

The town's agricultural roots are still evident in the fuzzy borders between town and country. A few barns stand inside the town limits; one resident on the southeast edge of town keeps some sheep in a small pen, and it's an easy five-minute walk to cornfields and pastureland from anywhere in Fullerton.

But farming in Nance County isn't what it used to be. Dan Willets, who was president of the Fullerton Area Chamber of Commerce in 1997, said, "The days are gone when the family farm supported the economy. Every year you have one more family that goes out of farming. Somebody else buys it and gets bigger."[2]

The median household income in Nance County was about $29,500 in 1997, and many of Fullerton's residents worked miles away in Columbus, Central City, Lindsay, or Albion, where there were jobs in manufacturing, sales, or government. So when Bell Family Farms—one of the nation's biggest pork producers—began looking for land enough to build fifteen confinement operations to produce half a million hogs a year in Nance County, Fullerton business people welcomed the news. Tony Lesiak, vice president of First National Bank, said, "I see a lot of young couples struggling and needing off-farm income, and with the demise of the small hog producer I was hoping some of our small producers could become a part of the Bell operation."[3] Lesiak thought Bell's offer to let neighboring farmers fertilize their crops with waste from the hog operations would be a plus.

Like Lesiak, Dan Willets anticipated new jobs and spinoff benefits for other businesses. "They were talking seventy-five to a hundred jobs, and those people would have to eat and they'd have to buy gas," said Willets. "And Bell said they'd try to use local corn and supplies."[4]

But to move into Nance County, Bell Farms needed willing sellers. The company was offering up to seven times the market price for unirrigated land. Emil Dubas, then sixty-seven years old, had always lived and farmed within four miles of the Nance County farm where he grew up. When he was approached about selling his land, he didn't take long to decline the offer, because Sand Livestock already had a fourteen-hundred-head farrowing operation about two miles from his farm. "It doesn't always smell," said Dubas. "But sometimes it's just terrible. It makes your eyes water and takes your breath away."[5]

Farmers wryly refer to the odor that emanates from hog lots as "the smell of money" because on a conventional diversified farm, pigs once provided the most reliable profits of all farm commodities and paid the day-to-day bills. For that benefit, farm families put up with occasional odor from a small pig operation, and it usually doesn't bother the neighbors.

But the stench generated by thousands of hogs can travel several miles. By 1993 scientists at the University of Iowa had established a connection between exposure to dust and toxic gases such as hydrogen sulfide and chronic symptoms experienced by confinement workers, including cough, scratchy throat, runny nose, burning or watering eyes, headaches, shortness of breath, and muscle aches and pains.[6] A 1997 Iowa study found similar ailments among people living within two miles of a four-thousand-sow confinement operation.[7] The National Pork Producers Council sponsors research and training to help reduce odor from hog operations, but the NPPC says, "While odors from pork operations may occasionally be distracting or irritating, they do not pose a health risk."[8]

Jim and Carolyn Knopik opposed the Bell plan because of their own 1970s experience with a five-hundred-sow confinement operation on their farm west of Fullerton. "I knew what it took to run one, and it's against my farming instinct," said Jim Knopik. "The hogs are crabby in those pens and more of them die."[9]

Diseases can spread easily among thousands of animals crowded together. Pigs can die of respiratory diseases, scours, erysipelas, pseudorabies, and hog cholera. Confined swine are routinely dosed with antibiotics—both to suppress disease and to promote growth. Many scientists fear that the common use of antibiotics in livestock that aren't sick is contributing to the rise of antibiotic-resistant bacteria, with profound implications for human health.

In spite of the heavy use of antibiotics in swine herds, there is inevitably a daunting volume of dead pigs to deal with. A thousand-sow farrow-to-finish farm typically produces about twenty tons of dead pigs a year.[10] Safely disposing of so many carcasses has become a growing problem in the industry.

Dead pigs were one of many concerns that Annette and Ron Dubas had about Bell's plans for Nance County. The Dubases, who farm about two miles south of the Knopiks, also rejected Bell's overtures. "We knew what they wanted to do wasn't farming," said Annette Dubas. "And they were a threat to my family. I was thinking about the future for my kids."[11]

The Dubases believed that big operators like Bell Farms would monopolize local markets and in effect deny small farmers a place to sell their hogs. They objected to farmland being acquired by investors and operated by wage-earning managers. Emil Dubas objected to "outsiders" coming into the neighborhood. "I wouldn't really care if a neighbor done it and lived there but they don't live there," said Dubas. "They're just a bunch of investors. Bottom line is all they're interested in."[12]

There's a persistent belief among opponents of the industrialization of agriculture that family farmers who provide the capital and day-to-day labor on property they own will take more responsibility for protecting the environment than nonresident investors concerned with profits more than with their children's future on the farm.

The American love affair with the family farm has deep roots. Thomas Jefferson, a farmer himself, wrote, "Those who labor in the earth are the chosen people of God, if ever He had a chosen people, whose breasts He has made His peculiar deposit for substantial and genuine virtue."[13] Big companies that bear little resemblance to the traditional family farm recognize the power of that mystique and use it to sell their own corporate image. For example, the nation's biggest pork producer in 1997 was Murphy Family Farms, which owned nearly three hundred thousand sows. The company was headquartered in North Carolina but did most of its business with contract growers in five states.

In its 1997 list of the top fifty pork producers in the nation, *Successful Farming* ranked Bell Family Farms as the twenty-fourth largest. The company owned twenty thousand sows and had operations in Colorado, the Dakotas, Minnesota, and Nebraska.[14] But owner Rich Bell argued that his company was very much like the other farms in Nance County. "We want to grow the family farm," he said. "We have been a family farm since 1849."[15] He said Bell Farms had a long, honorable tradition and that local people had nothing to fear.

Bell hosted an invitation-only meeting in Fullerton at the First National Bank's community room to win residents to his side. The banker Tony Lesiak, who has a degree in animal science from the University of Nebraska, came away feeling that Bell would run a clean operation and bring money into the county. But other considerations started to weigh on his mind. Fullerton residents were the main supporters of the Bell Farms plan, but Lesiak said, "People who live in town were never going to smell this thing."[16] Indeed, farmers in western Nance County were angry that business people in Fullerton were making plans for the future of the county without consulting those who would be most directly affected by odor from the hog farms. In protest, Emil Dubas pulled his money out of the First National Bank, where he had done business all his life.

Rumors spread about what half a million pigs in a sixty-square-mile area would do to the land, water, and air. To educate themselves on confinement issues, the Dubases and Knopiks collected stacks of information about what was happening elsewhere—much of it from the Internet. They communicated by phone and e-mail with more experienced activist groups who opposed factory hog farms in other states, including Wright County, Iowa, where hogs outnumber people twenty-five to one.[17] On 11 June Mid-Nebraska PRIDE, standing for "People Responding in Defense of Our Environment," was officially organized. Its leaders were Annette Dubas, Jim Knopik, and Ron Schooley, an organic farmer who lived about a mile from a site in Greeley County where Progressive Swine Technologies planned to feed more than fifty thousand hogs.

A few days later about two hundred people from Nance, Howard, and Greeley Counties filled the machine shop at Emil Dubas's farm to hear what PRIDE organizers had to say. After reading university studies, newspaper accounts, federal reports, and testimony from people involved in the issue in other states, the PRIDE organizers told the crowd that they had found nothing to convince them that raising half a million hogs a year in Nance County would be a good thing. The Knopiks had set up a rented copy machine in their living room and reproduced hundreds of pages of the material they'd collected to pass out at the meeting.

Mid-Nebraska PRIDE next met with the Fullerton Area Chamber of Commerce to explain their opposition to the Bell Farms plan. It was a tense meeting. But farmers and townfolk finally agreed they could cooperate both to protect the environment and to improve the area's economy.[18]

Business leaders like Tony Lesiak and Dan Willets reconsidered their enthusiasm for the hog farms. "I live on a farm with high nitrate content in the water, which is attributed to high use of nitrogen fertilizer," said Lesiak. "And we don't want to further contaminate that land with hog waste."[19]

Dan Willets read PRIDE's information and decided that earthen waste lagoons posed a threat. "They may or may not be leaking, but somebody needs to put a little more effort in to find out," said Willets. "Because if they aren't leaking, that takes away one of the arguments against the business; if they are, something needs to be done about it."[20]

Opposition to Bell's plans eventually prevented the company from finding enough people willing to sell land for its planned operations in Nance County. "When they saw we got organized and were putting up a stink," said Emil Dubas, "most of them backed out."[21]

In early July, when the Nance County Board of Supervisors took the first steps toward zoning, Rich Bell turned his attentions elsewhere—to unzoned Holt and Chase Counties, to the Winnebago Reservation in eastern Nebraska, and to the Rosebud Sioux Reservation in South Dakota.

Bell Farms quickly acquired an option to buy a 2,240-acre ranch in northern Holt County. The company planned to build a fourteen-thousand-head nursery and finishing units for thirty-six thousand pigs. On 23 July the Holt County Board of Supervisors—which had previously considered and rejected zoning—formed a committee to begin the zoning process.[22]

By late summer the Nebraska Department of Environmental Quality was reviewing proposals for about twenty new hog farms of at least a thousand hogs each in Boone, Clay, Dixon, Furnas, Greeley, Knox, Pierce, Seward, Harlan, Stanton, and Webster Counties.[23]

News of Mid-Nebraska PRIDE's efforts in Nance County had spread to other communities where hog farms were planned, and PRIDE leaders were invited to tell concerned citizens what they had learned. Even though their own fears for Nance County had passed for the moment, Jim and Carolyn Knopik kept the copy machine running in their living room. Billing themselves as "PRIDE Educators," Annette Dubas, Jim Knopik, and Ron Schooley took to the road every week that summer to counties in the eastern two-thirds of the state. At public meetings they explained what they believed to be the environmental, social, and economic dangers of mega–hog farms. Their efforts had an effect.

County board members, state senators, and members of Congress were swamped with phone calls and letters asking for a moratorium on new and expanded hog operations until their impacts could be studied and more tightly regulated. A moratorium would also give counties time to adopt zoning regulations. In mid-July about two hundred people from a dozen counties—mostly farmers and their families—traveled by bus and car to Lincoln to meet with state senators and environmental officials to demand a two-year moratorium on new hog confinements.[24] The Nebraska Pork Producers Association said such a measure would kill hog production in the state.

In nearly twenty years in the legislature, Lincoln senator Chris Beutler said he had never seen such a volatile and important issue develop so quickly.[25] Members of the legislature would have a steep learning curve if they were to gain enough knowledge of the issues to develop sound public policy.

On 15 August U.S. senator Bob Kerrey and six state senators held a three-hour public meeting at Columbus High School to hear concerns from a crowd of about five hundred people. Jim Pillen told the panel that "the 'hysteria of a few individuals' acting on the basis of 'fiction and sensationalism' was threatening major economic development opportunities in Nebraska."[26] Some among the crowd booed when Pillen said his operation near Wolbach was so well managed there would be no significant odor problems; others applauded and cheered his remarks.

Senator Kerrey suggested that the Clean Water Act could be used to protect the environment from pollution by big hog farms.[27] In fact, the Natural Resources Defense Council, an environmental advocacy group in Washington DC, had already sued the EPA to try to force the agency to use its authority under the Clean Water Act to more closely regulate agricultural runoff nationwide. But federal intervention would be years away. Meanwhile, state and local governments were on their own.

That fall the Nebraska legislature's Natural Resources and Agriculture Committees set out "to study the best way to locally control and encourage the growth and development of livestock operations considering the environmental concerns that go along with them."[28] The senators held public hearings in Fullerton, Bloomfield, and Crete. More than two hundred people packed each hearing, representing most of the voices that would be heard from again and again as state officials struggled with the issue for the next four years.

Randolph Wood, director of the Nebraska Department of Environmental Quality, described the agency's oversight of livestock-feeding operations. He said five people were responsible for permitting new operations and for inspecting about 1,400 operations with permits. Annually, the agency tried to inspect the 225 permitted operations that had more than two thousand animals.

But under questioning from Hastings senator Ardyce Bohlke, Wood admitted that the inspections seldom occurred and were even less likely with the flood of requests for construction permits the agency was receiving.[29] Because so few employees were assigned to do all the permitting and inspections, the NDEQ had a backlog of applications for construction permits. Winter was coming on, increasing pressure on the agency to approve permits.

It is interesting to read the 1997 hearing transcripts in light of later, more accurate information from the NDEQ. In 1999 the NDEQ would conclude that there were twenty-five thousand to thirty thousand hog- and cattle-feeding operations in the state, most of which had never applied for permits from the agency. State law had required permits since 1972. In the late summer of 1997 the NDEQ was apparently unaware of thousands of livestock operations in the state.

Although NDEQ regulations allowed lagoons to leak at the rate of one-quarter inch per acre of surface area, Wood assured the committee that properly built hog-waste lagoons would protect the groundwater. But if contamination reached the groundwater, Wood said it could be cleaned up. He said irrigators could use nitrate-polluted groundwater to supplement their crops' need for nitrogen fertilizer. Wood was unclear about how closely the NDEQ monitored the amount of lagoon waste used on crops to be sure that hog producers applied only the amount necessary to provide adequate nitrogen and phosphorus.

Wood also said the NDEQ had no authority to deal with "nuisances" like odor, which were best regulated by county zoning ordinances. Senators learned that livestock operations received NDEQ permits for the life of the facilities; in contrast, municipal and industrial wastewater treatment plants were required to renew their permits every five years, which provided for periodic NDEQ oversight.

Most of those testifying at the hearings acknowledged the need to pump economic life into rural communities, but they differed dramatically on how to do it. In 1997 several towns were competing to become the site for a new state prison. Omaha senator Ernie Chambers asked Jim Pillen whether pork or a prison would provide the most viable economic base. In a moment of unintended irony, Pillen answered, "Obviously I believe that the pork industry has much more to offer because it's going to use all our natural resources."[30]

While senators considered how best to protect the environment from a mountain of hog waste—an estimated six and a half million tons in 1997 alone—they were also asked to help preserve family farms.[31] How to do it was one point among many on which the state's two largest farm organizations differed. The Farm Bureau predicted a bleak future for rural Nebraska if limits were placed on confinement operations. The Farmers Union said big hog farms under contract to packers unfairly limited small farmers' access to markets and fair prices.

One small farmer said he wanted to be left alone. Lowell Stone of St. Edward said he'd been raising hogs since he was ten on a farm homesteaded by his family in 1883; with only 120 sows, he was easily able to compete with big operators. Calling himself "one of the little guys," Stone said he was concerned about his own future if the legislature were to limit his opportunities for expansion. "You're destroying my incentive to be a successful farmer if you accept this kind of thinking," said Stone. "Senators . . . with all due respect, I don't want your help, I don't want anyone else's help either if their goal is to keep me small and poor." He said, "I want to . . . work and learn from the larger producers in this business and to become successful and make a good life for my family."[32]

Susan Arp, a partner in Hidden Valley Pork near Albion, explained that twenty local families had cooperated to build a twenty-five-hundred-sow farrowing site. They would finish piglets on their own family farms. "Maintaining the individuality of our family farm is important," said Arp, "but we cannot overlook the opportunities, changes, and potential provided by joining together with others in a large project."[33]

One of Nebraska's biggest hog producers—Sand Livestock—was represented by its chief financial officer, Tim Cumberland. He said that with thirty years of experience in the hog business, the company had a good environmental record. He said Sand paid an average salary of over $26,000—and slowed the population loss from rural communities.[34] A Sand employee, Jeff Lindgren, said his job made it possible for him to stay in the Albion area and farm on the side. "I was able to make my dream come true by working for Sand Systems," said Lindgren. "I've been able to support my family nicely and (give) them an opportunity to grow up on a family farm."[35]

Jim Pillen said misinformed critics of modern hog production were spreading "hysteria" around the state. "This is our home too," said Pillen, "and we want to keep it clean for our families and our neighbors."[36]

Unconvinced, Ron Schooley renewed the call for a two-year moratorium on new hog facilities. "If these things are as wonderful as Dr. Pillen and everybody tells us they are, they'll be just as wonderful in two years as they are now," said Schooley. "If they are as dangerous as we believe they are, we need to know now. We don't need to wait two years and then find out we've made a horrible mistake."[37]

Neighbors of operations run by Pillen, Sand, and other producers told about infestations of flies from nearby hog confinements, hog carcasses lying by the road for days until rendering companies could pick them up, and lagoon waste spilling into county road ditches and farm ponds. They accused the NDEQ of indifference to these problems.

Annette Dubas reminded the senators that 85 percent of Nebraska's drinking water comes from groundwater. She said the waste from thousands of hogs would put too great a burden on the ecosystem.[38]

University of Nebraska–Lincoln scientists tried to ease fears about air and water pollution from big hog farms. They cited studies showing that manure seals lagoons and reduces the chance of leaks. They assured senators that experts in swine management at UNL and Iowa State were doing research to control odor.

At the Crete hearing in October, the senators first heard about "dead pits" used by Sand Livestock to "liquefy" thousands of tons of hog carcasses each year. The liquefied carcasses were then spread on cropground. The process wasn't included in a state law that set out legal methods of disposing of animal carcasses.

Notable by its absence from the hearings was any testimony from Premium Farms, a company that had generated public concerns in northeast Nebraska—particularly in Antelope County where one man took the law into his own hands. Brunswick farmer Rich Wells had tried to persuade construction workers on the Premium Farms site across the road from his farm to prevent soil eroding from the site onto his land and to settle the construction dust that made him and his livestock ill and clogged his center-pivot irrigation system and corn dryer. The construction crew ignored his pleas. Wells complained to county officials who couldn't intervene because the county had no zoning ordinances.

On the night of 17 October, Wells, a hired hand, and a friend vandalized the construction equipment; they fired bullets through radiators and fuel tanks, broke windows, and punctured tires on trucks and bulldozers.[39] Wells was arrested and convicted of three counts of criminal mischief. Evidence presented at trial showed that Premium Farms had built on the site across from Wells's farm and seven other sites without construction permits from the NDEQ. The agency sent a letter to the company warning of the violation but levied no fines or other penalties.[40] Wells—a formerly law-abiding citizen who now had a felony record—was doing time in the Hastings Correctional Center when fourteen thousand pigs moved into new barns across from his farm in Antelope County.

By early fall of 1997 the NDEQ was receiving forty-five to fifty phone calls a day regarding livestock operations. Callers included permit applicants, people concerned about environmental problems, bankers asking about the status of permits, and local government officials and attorneys asking about permitting procedures.[41]

In December Nebraska governor Ben Nelson gathered representatives of the livestock industry for a capitol press conference. He announced that Randolph Wood of the NDEQ and State Director of Agriculture Larry Sitzman would head a working group to report to him on the livestock waste controversy. Nelson wanted to know, "What is the science of this area and what kind of technical responses can we develop to make sure that agriculture is not undermined nor is the environment?"[42]

Anticipating that the legislature would respond to public demands for action, Sitzman said, "Environmental regulations must not restrict, either economically or environmentally, agricultural growth," and he pledged to bring to the effort "sound scientific data rather than emotional rhetoric."[43]

The seventeen members of the panel included representatives of the Farm Bureau, Farmers Union, Nebraska Cattlemen, Pork Producers, Corn Growers, the Nebraska Chamber of Commerce, Mid-Nebraska PRIDE, and the natural resources districts. They were to report to the governor by 10 January 1998, one month after they were appointed.

In 1997 citizens in at least fifteen states were trying to come to terms with the environmental, economic, and social consequences of rapid growth in the hog confinement industry. To give policy makers time to consider how best to respond to the trend, the National Catholic Rural Life Conference in December called for an immediate nationwide "moratorium on the expansion and building of new farm factories and . . . for a serious consideration of their replacement by sustainable agricultural systems which are environmentally safe, economically viable, and socially just."[44]

The statement received little media exposure in Nebraska, but Catholic residents of Nance County took note. In coming months, a Catholic priest would often join the leaders of Mid-Nebraska PRIDE in pleading for caution from policy makers. Big hog operators continued to break ground for new confinement units across the state.

Every facet of local government has a profound idea behind it.
It involves money, power, politics, class, race. It's about who can
afford to move into your town and who can't, about who's going
to get rich developing their land and who isn't; it's about why
they never build the sewer plant in the rich part of town. It's
about what people are willing to spend their tax money on and
what they are not willing to pay for. It's about what people care
about and how they want to live.
– Bruce DeSilva, National News/Features Editor with the Associated Press, from a speech to the National Writers Workshop,
Omaha, Nebraska, 8 May 2000

Home Rule

While the Nebraska legislature and the governor studied how best to respond
to the influx of swine operations, local officials were feeling intense pressure
from constituents—particularly in about a dozen unzoned counties that
had been targeted by big hog producers. In Harlan County, two small towns
found a way to respond more quickly than their county board, taking action
that resulted in years of litigation over an important state constitutional
issue.

In the Republican River Valley, water has been both a treasure and a mur-
derous force. In the dry early spring of 1935—after a year of drought—high
winds kicked up blinding dust storms across the region, adding to the misery
of the Great Depression. Then in early May, rain came in torrents. By 20
May tributaries of the Republican River were flooding. On Memorial Day
a wall of water swept down the Republican Valley for nearly two hundred
miles, wiping out farmsteads, killing at least ninety-four people and thousands of head of livestock, and causing about ten million dollars in property
damage.[1]

Over the next twenty years the U.S. Army Corps of Engineers spent millions to build five dams to contain the Republican River and to provide irrigation water for about ninety thousand acres of farmground.[2] Harlan County Lake alone provided 104 billion gallons of storage to control flooding and to irrigate thirty-five thousand acres of rich farmland in Nebraska and Kansas. Like the other new reservoirs on the Republican River, Harlan County Lake became a magnet for recreation, spawning tackle and bait shops, marinas, motels, and campgrounds—especially in Alma, a town of about twelve hundred on the north shore of the lake.

By the 1990s an area once called "Death Valley" in the aftermath of the 1935 flood was lush with irrigated crops and teemed with recreational traffic as people flocked to Harlan County Lake for fishing, waterskiing, and boating.

But like most of rural Nebraska at the end of the twentieth century, Alma, the Harlan County seat, was losing population and looking for new ways to sustain the economy. Then in the summer of 1997, Furnas County Farms applied to the Nebraska Department of Environmental Quality for a permit to build a 34,000-hog farm about eight miles from Harlan County Lake. The facility would be built by Sand Livestock Systems, Inc., at a cost of four or five million dollars and would finish about 80,000 hogs a year.[3] Each year it would provide wages of $225,000 for ten employees and would consume about a million bushels of corn and pay $45,000 in property taxes.[4]

Corn and cattle prices were down in 1997, and Harlan County had lost two-thirds of its hog producers since 1992. Only eighteen farms still raised pigs in 1997, and there were only about 7,500 head of pigs in the county, compared to 13,485 on fifty-three farms in 1992.[5] So some residents welcomed the news that Furnas County Farms wanted to bring 34,000 hogs into Harlan County. Farmer Ken Ohrt, who sold some land to the hog company, said, "We're not gonna get an IBM or computer chip factory in this area for people to work in, and so we have to go with what we've got, and this is agriculture."[6] Ohrt anticipated saving money on commercial fertilizer by allowing hog waste to be applied to his cropground through irrigation pivots.

But Tom Thomas, the mayor of Orleans—a town of about five hundred people on the north bank of the river about ten miles upstream from Alma—saw unacceptable costs that canceled out the economic benefits promised by the hog farm. "When you take 30,000 hogs, put them on one small piece of ground, and put all that manure and waste into lagoons over strata that will allow it to percolate into the groundwater, you're going to have problems," said Thomas.[7]

The proposed swine operation would have thirty-one buildings. Three waste lagoons with a total surface area of twenty-six acres would hold 145 million gallons of hog waste. The site was about three miles from where Orleans planned to dig new drinking-water wells.[8] They would replace old wells that were heavily contaminated with nitrate from commercial fertilizer. The hog-farm site was also five and a half miles from wells that provided Alma with drinking water.

Residents of Harlan County began to put pressure on the Alma City Council. Mayor Hal Haecker said, "They were asking us why we weren't concerned about our groundwater, drinking water, and the lake."[9] But there was apparently little that town governments could do to limit development outside their municipal boundaries.

In 1997 only about a third of Nebraska's ninety-three counties were zoned. Like so many other counties, Harlan County held to the long-cherished premise that what people do with their own land is their business. But when rumors spread of plans for the hog farm, there was a sudden demand for zoning. More than four hundred people opposing the development signed a petition to the Harlan County Board of Supervisors on the day the board voted to start the countywide zoning process.[10] Until the comprehensive plan and zoning regulations were adopted—a process that wouldn't be complete until 2001—the hog farm needed only the blessing of the Nebraska Department of Environmental Quality.

Some citizens weren't convinced that their water would be protected by NDEQ regulations that allowed earthen lagoons to seep at the rate of one-quarter inch per day, and that permitted center-pivot application of the waste onto highly erodable sloping ground. Despite assurances from Furnas County Farms and Sand Livestock officials that their "state-of-the-art" operation would be safe, Pam Pape, who lived three miles from the swine farm site, feared for her family's health. "I'm concerned with our groundwater and drinking water and concerned about the pivots putting this stuff on," said Pape. "I'm concerned about the viruses and disease in this water that will be put into the air that the wind will blow that we all will breathe."[11]

In the absence of zoning, the Harlan County Board of Supervisors took the only steps they could to restrict the hog farm. The board denied easements for the company to run pipes under county roads to transport hog waste from lagoons to center pivots. In a move marked both by urgency and a sense of humor, two township boards voted to prohibit removing animal waste from confinement buildings and depositing it within the township.[12] The ordinances were similar to those passed by townships in Nance, Antelope, and Holt Counties in efforts to discourage hog operations there.

Residents who were alarmed by the prospects of hog waste polluting both the groundwater and Harlan County Lake wrote letters to local papers objecting to the proposed location and size of the operation.

The Orleans Village Board and the Alma City Council together hired a Kansas City engineering consultant to evaluate the environmental impact of the hog farm's design and location. The consultant found that the clay-lined waste lagoons would be built on porous soil. He said the liquid waste to be spread by irrigation pivots on surrounding cropground would drain into waterways flowing into Harlan County Lake. The consultant said hog waste from the operation would reach the groundwater within three to five years. It would reach Alma's wells within ninety-seven years and Orleans' wells in about fifty years.[13]

Some residents were unconcerned, but Orleans' Mayor Thomas said most people felt an obligation to future generations. "If we don't take responsibility for it now," said Thomas, "there's no way you can turn back and clean that water up once it's contaminated."[14] Alma city councilman Rick Calkins said, "We felt like, as a council, that even if it was a hundred years down the road, we could not leave a legacy of contamination a hundred years down the road."[15]

The Alma City Council acted first, relying upon an obscure state law that allowed small cities to protect their drinking water within fifteen miles of the city limits. Following the recommendations of the Kansas City consultant, in October the council passed ordinances requiring any operation with more than twenty-five hundred hogs to:
- apply to the city for a permit before beginning to build;
- line its waste lagoons with a seep-proof synthetic liner laid on top of packed clay or use concrete or steel tanks to contain the waste;
- apply manure to cropground only to the extent that the nitrogen, phosphorous, and other nutrients could be used by growing crops;
- install groundwater monitoring wells around the site; and
- provide a bond so the operation could be closed without cost to the taxpayers.[16]

The ordinances were much more restrictive than state law. An attorney for Furnas County Farms said the city had no authority to pass them and that the company would build as planned.[17] Furnas County Farms sent its own geotechnical report to Alma, along with another letter from their attorney, who wrote, "We are totally convinced that this project will not negatively impact the environment of Harlan County in any way. We think we will be a good citizen of your county."[18]

On 1 December the NDEQ granted a construction permit. Without permission from the city and with a state permit in hand, Furnas County Farms and Sand Livestock began to build the hog facility.

Orleans had quickly passed ordinances similar to those passed by the Alma City Council, but Alma ended up in court. Alma asked the Harlan County District Court to compel Furnas County Farms to apply for the permits required by the new ordinances. Furnas County Farms responded by saying that the ordinances were unreasonable, violated state law and the constitution by exceeding the city council's legal authority, and were designed solely to prohibit the hog farm from being built in Harlan County. In a counterclaim, the company asked for damages of four million dollars to cover business losses caused by the delay in building its thirty-four-thousand-hog operation.[19]

Apparently to avoid regulation by the city, the company had reduced the size of the operation to twenty-two hundred hogs—below the level that required a permit from the city. Gary Gausman, president of Sand Livestock Systems, said the twenty-two-hundred-head operation would generate only a tenth of the property taxes and one-tenth of the jobs promised with the thirty-four-thousand-head facility.[20] Although the hog companies had seemed to comply with Alma's ordinance, they nevertheless pursued the lawsuit challenging the city's authority.

The case would reach the Nebraska Supreme Court twice. While the wheels of litigation slowly turned, two other small towns passed similar ordinances—Long Pine and Bassett in north-central Nebraska, where Premium Farms wanted to confine a total of eighty-four thousand hogs on six sites in the watershed of Long Pine Creek, one of Nebraska's few trout streams.

By fall 2002 the Nebraska Supreme Court had not made a final ruling on the Alma case. A decision either way is expected to settle an important state constitutional question, whether the state has the sole authority to regulate livestock operations and what, if any, authority rests with local governments.

Terry Woollen, who farms near Alma, was concerned about the impact a big hog operation could have on the area's water. But he also wondered whether, if Alma's ordinance stood up in court, townspeople would make other attempts to tell farmers how to operate. He said farmers were apprehensive about a town government having any control over them, and yet some farmers strongly opposed the hog farm. Woollen said, "It's one of these things where people do not like government intrusion until they have a problem and then they want government to be an intercessor on their behalf."[21]

With exactly that change of heart in evidence, many counties that had previously rejected zoning hastened to put it in place. In Boone County, where neighbors of two hog operations owned by Progressive Swine Technologies suffered from the odor they emitted, the county board of supervisors felt pulled in two ways. Some constituents pressured the board to support hog confinements as economic development; others demanded that they do everything possible to keep the operations out of the county.[22] County officials, citizens, pork producers, and other livestock and business interests would all look to the legislature to resolve the conflict between opportunities for profit and threats to the environment and quality of life in rural Nebraska.

Historically, education and property taxes were the issues most likely to generate legislative debate over the merits of local control. The Nebraska legislature typically came down on the side of local control. But sentiment shifted when the 1998 legislature considered Legislative Bill 1152, Ord senator Jerry Schmitt's proposal to allow counties to enact emergency zoning when faced with sudden, unanticipated development. The bill would have authorized counties to adopt short-term regulations—so-called interim zoning—to control rapid new development while they worked on comprehensive plans and zoning ordinances. LB1152 would have created an incentive for counties to go through the entire zoning process; they could enact interim zoning only once in five years and it would be valid for only two years.

Senator Schmitt was a burly, retired thirty-year veteran of the State Patrol in his second term representing the rural Forty-first District in central Nebraska. Schmitt figured his constituents had elected him to office largely because of his support for the death penalty—over an incumbent who supported repeal—and because of his opposition to a mandatory seat-belt law that many Nebraskans believed infringed on individual freedom.[23]

But in supporting interim zoning, Schmitt came down on the side of more rather than less government regulation. In early February, at the standing-room-only hearing on LB1152, Schmitt said big hog operations created unacceptable risks in unzoned counties. Without zoning, he said, "There's no way to regulate where they locate their lagoons or how close they build to neighbors or towns or water sources. There's no way to limit the amount of livestock they bring in. In short, there's no way to protect the quality of life for farmers and ranchers already operating in a county or the townspeople who live there."[24]

Schmitt's proposal came up against a potent team of lobbyists for the Nebraska Pork Producers Association, the Nebraska Cattlemen, poultry growers, the Farm Bureau, and the Nebraska Chamber of Commerce and

Industry. Their message was straightforward: interim zoning would, in effect, shut down the state's livestock industry for two years and was just a backdoor way of enacting a moratorium.

Hog-farm opponents had, in fact, continued pressing for a two-year statewide moratorium to give counties time to adopt zoning and to give the state time to study how best to regulate intensive livestock feeding. Attorney General Don Stenberg issued an opinion saying a moratorium could be based on the legislature's police power, that is, the power "to protect public health, safety and welfare, as well as the state's environmental resources."[25] He said such a measure would have to be carefully written to avoid a constitutional challenge. But it was clear that the legislature didn't want to consider a statewide moratorium. That left any kind of timely response up to local governments—the reason for Senator Schmitt's bill.

At the hearing on LB1152, the twenty people speaking in its favor included individual farmers, county officials, and representatives of the Farmers Union, the Audubon Society, and the Sierra Club. Although county officials reassured senators that LB1152 wouldn't stop livestock development, the "M" word peppered opponents' testimony. The president of the Nebraska Farm Bureau foretold a gloomy future for the state's economy if its chief agricultural enterprise—livestock feeding—were to be subjected to a moratorium on growth. The lobbyist for the Nebraska Chamber of Commerce and Industry feared that counties would use interim zoning to halt all new business development for two years.[26]

Many critics of the bill said interim zoning would open the way to emotional rather than science-based decisions about livestock feeding. A Farm Bureau lobbyist, who objected to the emotional tenor of local debates over hog farms, said, "We are for local control, but we are for responsible local control where there's rational discussion of the development of livestock operations."[27] Ron Schooley of PRIDE was insulted by the implication that farmers—who made up the majority of rural county board members in the state—wouldn't inform themselves and act wisely and logically to protect human health, the environment, and agriculture.[28]

As debate opened on the bill in March, the capitol rotunda was filled with paid lobbyists from livestock and business groups opposing the bill and with unpaid grassroots activists who came from around the state to support it. Although supporters of interim zoning had outnumbered opponents more than two to one at the hearing, and although most supporters of LB1152 were farmers and county officials, there was little sympathy for unzoned counties in the 1998 legislature.

Nebraska's counties had had the authority to zone agricultural land for about thirty years, and many senators criticized them for failing to anticipate the growth of hog farms. But a handful of others argued that, like individuals, government bodies—including state legislatures—are typically disinclined to take preventive action until threatened.

Among those most sympathetic to the demands for interim zoning was Omaha senator Ernie Chambers, a twenty-eight-year veteran of the legislature, who identified his occupation as "defender of the downtrodden."[29] Chambers said senators had an obligation to provide a remedy for unzoned counties faced with unexpected development, just as a doctor would help a sick or injured person who could have taken precautions to prevent a disaster.[30] To encourage empathy in the legislature, another Omaha senator, Don Preister, suggested, "Whenever hog confinement legislation is being discussed, an open pail of hog manure should be placed in the center of the Assembly and a fan be located so as to distribute the odors throughout."[31]

But concerns for quality of life fell on deaf ears. Senator Bud Robinson, who represented Cuming County, which had the state's highest density of livestock producers, said odor was "the sweet smell of success."[32]

Many rural senators—most of them farmers—said county boards, given the option of interim zoning, would be too easily swayed by emotion and pressure from local constituents trying to stop big livestock developments. Senator Roger Wehrbein, a Plattsmouth farmer and the influential chairman of the appropriations committee, said he was resigned to a future he described as the "Wal-Marting of agriculture," as big operations with greater margins of profit replaced traditional family farms.[33] As much as some Nebraskans might want to protect family farms, Wahoo senator Curt Bromm, an attorney, said, "I don't believe in advancing a socioeconomic position with a zoning policy or with an environmental policy."[34]

Others would say the legislature did just that in failing to give counties the authority for emergency zoning in 1998. Senator Schmitt withdrew LB1152 after the legislature amended it—at the urging of livestock and business lobbyists—to dictate the interim zoning rules that a county could adopt. The amendment essentially removed the possibility that local governments could shape the interim zoning rules to their needs.

Lincoln senator David Landis accused the legislature of being "deeply schizophrenic on the topic of local control." He said, "We are for local control when it suits us. And then we're not for local control when we don't want it." By failing to pass LB1152, Landis said the legislature ignored the pleas of Nebraskans who said the law didn't allow them to "act quickly enough to meet a real threat that we think we have in our jurisdiction."[35] Senator

Cap Dierks, who represented the Fortieth District, where Premium Farms was building so rapidly, had been listening to the fears and complaints of his constituents for months. When interim zoning died, he asked the legislature, "How am I going to tell my people? . . . They're sitting out there having these things built across the road from them and no one can stop them."[36]

After LB1152 died in the 1998 legislature, many Nebraska counties finally starting the zoning process as the pressure from hog-farm developers increased.

For those counties that still delayed, Sally Herrin of the Nebraska Farmers Union offered a vivid analogy: "Zoning is a lot like oral hygiene. Unless and until it's an issue in your own life, it's a real bore. But let the teeth start to loosen in your head, let the heavy machinery commence excavation on a hundred-acre hog waste lagoon in your county, and suddenly you find yourself getting real focused, real fast."[37]

In the spring of 1998, cattle, corn, and wheat still dominated the rural landscape of Nebraska's far southwest corner. In Dundy County, Bowman Family Farms applied for state permits for facilities to store the waste of more than five hundred thousand hogs. In Perkins County, Enterprise Partners—a company with connections to Sand Livestock—proposed building a nursery for about sixteen thousand pigs and a finishing site for twenty-five thousand. Sun Prairie, a partnership connected with Bell Farms of North Dakota, was preparing to lease land from two Chase County families to build facilities for about eighty-four thousand hogs.

The Imperial Chamber of Commerce cheered the prospects for new business and new property-tax revenue. The editor of the *Imperial Republican* visited a Bell Farms operation in Colorado and then promoted the development editorially.[38] One businessman from Ogallala, fifty miles to the north, anticipated twenty-four million dollars in new business coming to the community from the grain, labor, construction, trucking, and other services the hog farms would require.[39]

But not everyone was enthusiastic, and many residents called for the Chase, Perkins, and Dundy County boards to start the zoning process. Meanwhile, the controversy landed in the laps of the Upper Republican Natural Resources District's board of directors.

Nebraska's twenty-three natural resources districts—unique in the nation— are another example of the state's fondness for local control. As locally elected guardians of the state's water resources, the NRDS have a formidable task.

Nebraska rests upon about 67 percent of the Ogallala Aquifer, which historian John Opie has described as "one of the nation's great hidden treasures."[40] The aquifer contains more than three billion acre-feet of water (an acre-foot is 325,851 gallons—enough water to cover one acre with a foot of water). It stretches from northern Texas to South Dakota, but massive amounts of water drawn from the aquifer for irrigation and livestock are depleting it much faster than it is being recharged. The annual depletion of the Ogallala Aquifer is measured in feet while recharge is measured in inches.[41]

Confined hogs use a lot of water, both for drinking and flushing the barns—as much as forty-three gallons a day for a sow and her litter and twenty gallons for each finisher pig.[42] Millions of pigs are sucking up Ogallala water from South Dakota to Texas since the pork belt has expanded into the Great Plains.

Between 1980 and 1995 groundwater in the aquifer beneath the three counties of the Upper Republican NRD had declined as much as twenty feet.[43] In 1997 the eleven-member NRD board—most of them irrigating farmers—responded by banning the drilling of new irrigation wells, at a time before hog farms were on the horizon, when grain soaked up the irrigation water and cattle accounted for the majority of livestock. In the late 1990s the Upper Republican NRD's concern for the declining water table clashed with hog producers' arguments that they should be able to use water to grow pigs just as others were using it to grow crops.

Robert Ambrosek, an irrigator and a twenty-five-year member of the NRD board, struggled with the question of how to be fair to everyone. He recognized that many local residents looked to the NRD board to keep out big hog farms by restricting their access to water. But he couldn't justify that approach simply because one person wanted to use water to raise pigs and another wanted to use it to raise corn. "We have to look at the resource and manage it and get as far as we can and treat everyone as fairly as we can," said Ambrosek. "But any time you have restrictions, someone is going to be hurt."[44]

Threatened by a lawsuit, the NRD agreed to let the hog farms draw water from the aquifer using fifty-gallon-per-minute wells.[45] Under state law at the time, NRDS didn't have the authority to regulate fifty-gpm wells, which were typically used for domestic purposes. Three grain farmers who had been denied additional water for their crops sued for their own rights to the water.[46] The lawsuit has yet to be resolved.

As unzoned counties began the long process of drawing up comprehensive plans and adopting zoning regulations, citizens across the state demonstrated a mutual concern for protecting their air and water from pollution. Counties typically placed more restrictions on livestock operations than the legislature was willing to adopt, and those restrictions varied from county to county. Consequently, the state became a patchwork of rules, forcing livestock producers to adapt their methods of operation when they crossed county lines.

Some counties required manure to be injected into cropground rather than being applied through center-pivot irrigation. Some required that lagoons be covered to control odor and that synthetic liners be installed to limit seepage. Every Nebraska county that has adopted zoning has set a minimum distance between new livestock operations and homes or towns. Because rural counties lack the money and expertise to monitor air quality, these so-called setbacks, which range from a half mile to a couple of miles, were one way to protect citizens from the odor generated by thousands of head of livestock. To control both odor and runoff, Lincoln County—one of the nation's biggest cattle-feeding counties—prohibited center-pivot waste disposal and required manure to be contained above ground in concrete or metal containers rather than in the usual clay-lined pit.[47]

Counties have required manure management plans from livestock producers and have required them to apply for operating permits. The permitting process involves public hearings, giving local residents a voice in the process. Local zoning hearings were often emotional as new pork producers moved into a community or as long-time farmers tried to expand their operations.

There is an intensity to county board hearings that is seldom present at a hearing of the legislature. People serving on county boards and zoning commissions know personally most of those testifying at hearings. They are neighbors, customers, or members of the same church; their children go to school together and they are related by marriage. There were tense, crowded meetings in courthouses across the state as county officials struggled to shape ordinances that neighbors and friends who had elected them could live with. Setbacks were often the most contentious item; although many of Nebraska's rural counties are sparsely populated, requiring too great a distance between livestock and the nearest neighbors would mean little or no new livestock development in a county. But hog odor can travel several miles. Where very large hog farms have been built, the typical setbacks of a mile or so have done little to sweeten the air for neighbors.

Every county faced with growth in hog production had moments of drama as the county board tried to find compromises that citizens could live with. Young farmers who wanted to enlarge relatively small swine operations were suddenly forced to explain themselves to county boards who struggled with public sentiment opposed to any new swine operations, regardless of size.

In a six and one-half hour hearing in March 2001, the Polk County commissioners struggled to sort out conflicting testimony over a local family's plans to set their son up with a ninety-nine-hundred hog farm. Reflecting the pressure he was under, the commission chairman told the crowd, "We're trying to make a right decision. I don't know how we're going to make a popular decision."[48]

In Gage County it was rumored that Bell Farms wanted to build a massive swine farm in the county. At the same time, twenty-nine-year-old Rod Linsenmeyer—a fifth-generation Gage County farmer—planned to expand his farrow-to-finish operation. Unaccustomed to public scrutiny of his business, Linsenmeyer trembled as he told the county board and an overflow crowd of citizens how he would control the odor by disking most of the hog waste into the soil rather than disposing of it through center pivots. Facing intense pressure to deny any new hog operations in the county, the board took several weeks to approve Linsenmeyer's expansion.[49]

In Holt County long court battles had been the only recourse for residents living with the stench from National Farms. Kelly Huston, whose family had farmed without controversy in the county for forty years, met stiff opposition when he wanted to expand his operation from forty-eight hundred to ninety-six hundred feeder pigs.[50]

In Bassett about two hundred people sat through four hours of a hearing on Premium Farms' application to build six hog farms with fourteen thousand head each. They stood, cheering and applauding, after the three-member Rock County board turned down each permit, one by one.[51]

Cheers, hisses, and boos often interrupted normally sedate county board meetings as they took up divisive zoning issues. But in unzoned counties targeted by hog operations, zoning proposals—which had previously led to short political lives for many a county commissioner—were suddenly popular. Under the steady gaze of family, neighbors, business people, and friends, county commissioners statewide tried to sort out what would be fair. Like many state senators, members of county boards have been heard to say that they personally wouldn't want to live near a hog farm. And yet most of the time, contrary to what the critics of interim zoning expected, county boards voted to allow new or expanding hog operations, meaning someone would have to smell the pigs. The Nance County Planning Commission unanimously recommended denying a permit for Furnas County Farms to double a fourteen-hundred-hog operation. But the Nance County board voted to allow the permit, even after Emil Dubas, his voice hoarse with emotion, told them how bad the odor already was at his farm two miles from the operation.[52]

Tears were common among people testifying at local hearings on hog-farm permits. Most citizens are unaccustomed to being placed in the spotlight among dozens of their neighbors, and zoning hearings were typically packed with citizens who had strong feelings either for or against the permits.

As the controversy spread across the state, the Nebraska Pork Producers Association tried to represent members of all sizes of operations. Regarding some pork producers who complained about local restrictions, board member Dan Hodges wrote, "I feel it is extremely arrogant of anyone, livestock producers or industry, to feel that they are so important that they do exactly as they please in the name of efficiency and economic development, and the local people are just going to have to 'live with it.'"[53]

Local officials had to consider the consequences of limiting economic development at a time when the legislature had capped local property-tax levies. Increased property valuations were needed to make up the loss in tax revenues. But these practical considerations had to be balanced with people's concerns for quality of life. One consultant who guided several counties through the zoning process said zoning regulations were written quickly to respond to public pressure. Even in the spring of 2001, after zoning fever had hit and passed through most of Nebraska's rural counties, he said emotion continued to play a role in the process.[54]

Emotions are an important ingredient in sorting out public policy on any issue that affects people where they live. Anthropologist Kendall Thu of Northern Illinois University studies how food production and distribution in industrialized countries affect the social, political, and economic order and environmental and public health. Thu says the level of emotion over an issue "demonstrates the level of caring and regard people have for their homes, neighborhood, and community. The fact that the so-called scientific world is not emotional is more problematic to me. Only when you have that concern and regard do you seek to protect what's cherished."[55]

The haste with which some counties adopted zoning made them vulnerable to lawsuits. Premium Farms sued Holt County, alleging that the county board violated state law by regulating farm buildings. At issue was a state law that granted counties the power to zone land use but at the same time seemed to bar them from zoning farm buildings. In March 2002 the Nebraska Supreme Court unanimously ruled against Premium Farms in an opinion that endorsed broad powers for county zoning. The Court wrote, "[W]e are guided by the presumption that the Legislature intended a sensible, rather than an absurd, result in enacting the statute," and "Land use does not stop at the walls of a building. Instead, land use is inextricably interwoven with what occurs on the inside and the outside of buildings."[56]

Companies connected with Sand Livestock were particularly litigious in efforts to overturn local zoning regulations. First there was Furnas County Farms' legal dispute with Alma in Harlan County. Enterprise Partners successfully sued Perkins County for passing zoning regulations before adopting a comprehensive plan, as required by state law. In 2000, after Furnas County Farms began moving dirt for a forty-four-thousand-hog confinement in Hayes County, the county board of commissioners required waste lagoons to be covered and banned the land-application of manure. The company sued, alleging among other things that the regulations exceeded the county's authority under state law and were "unduly burdensome" and "not rationally related to . . . any legitimate governmental purpose." A federal judge disagreed on that point, but it would take nearly two years of legal maneuvering before Furnas County Farms dropped its complaints against Hayes County.[57]

The lawsuits in Nebraska echo similar disputes in other states where the courts are being asked to decide where regulatory authority lies—with the state or with local government.

Across the nation, counties and townships have typically chosen to regulate big hog farms more strictly than their state legislatures—with mixed results.

In 1999 the South Dakota Supreme Court upheld the right of a county board to consider "population, impact on county roads, devaluation of surrounding real estate, noxious odors, and pollution" in denying a permit to a six-thousand-hog operation.[58] But in the same year, the South Dakota Supreme Court ruled that a township lacked the authority to regulate a hog farm.[59]

A federal court in Michigan upheld a township ordinance that sought to limit the odor from farms by limiting the number of animals farmers could have on their property.[60] A Missouri court of appeals upheld county health ordinances designed to protect residents from health hazards associated with hog farms.[61] But in North Carolina the state supreme court struck down a county ordinance intended to protect the health of people living near hog farms. The court ruled that state law superseded local law in this respect and wrote that if counties were allowed to devise their own rules for hog farms, "The result could well be that the rights of adjacent landowners in each individual county would be substantially elevated above the rights of swine farmers to workable, nonexcessive regulations."[62]

After Humboldt County, Iowa, passed ordinances to protect air and water quality from big hog farms, the Iowa Supreme Court ruled that Iowa law preempts counties from regulating livestock operations, leaving that role to the Iowa Department of Natural Resources.[63] Bills to restore some authority to the counties stalled for several years in the Iowa General Assembly but were revived in 2002. The Kansas legislature has given counties the authority to permit or prohibit corporate farming within their borders.

In Nebraska the interim zoning idea would be revived in 1999 for a more receptive legislature. Although it appeared that no rural county would escape the controversy, by January 1999 twenty-two counties had still not begun zoning. It was clear that many counties would continue to delay until they attracted the notice of big pork producers looking for land and water.

Nebraska's patchwork of zoning ordinances forced operators to change their practices depending on where they were located in the state. It was predicted that livestock producers would go "pollution shopping," looking for counties with the least stringent requirements. That precise situation had developed on a national level as big hog producers, stymied by strict environmental regulation in some states, shopped for states with fewer restrictions. There were calls for the federal government to intervene because air and water don't recognize political boundaries, nor should the laws protecting them from pollution.

It would take years for the EPA to intervene. Meanwhile, in Nebraska— as in at least twenty other states—local and state policy makers were left to determine just how far to go in regulating a major sector of their farm economy.

I, for one, want to have clean water as much as anyone else.
[On] the other hand, I think we have to be realistic as to what
we can and cannot do and what we can afford and what we
can't afford to do.
– State Senator Roger Wehrbein, Debate Transcript, Legislative
Bill 1209, 30 March 1998

The Legislature Weighs In

In its sixty-day session the 1998 Nebraska legislature considered and passed
bills to reform the death penalty, increase penalties for drunk driving, provide property-tax relief, and create a sex-offender registry. In addition to
the hours of public testimony, negotiation, and debate on these weighty
issues and dozens of others, senators also spent an unprecedented amount
of time in 1998 considering hog waste: its chemical makeup; its gallons
of production per day per boar, sow, piglet, and market hog; its aromatic
qualities and how they arise in an anaerobic context and in the spray from
center-pivot irrigation; its value as fertilizer for corn, wheat, soybeans, and
alfalfa; the chances of its polluting the state's water or indirectly enriching
rural communities; and whether and how the state should regulate it.

A dominant theme during the many hours of testimony and debate over
these issues was a call for "sound science." Citizens and lobbyists rounded
up experts to support their own points of view. The legislature was flooded
with contradictory research and conflicting views of the future. Livestock
industry lobbyists, bankers, and chambers of commerce promised a lift for
sagging rural economies. Others predicted environmental disaster and the
death of family farms if big hog confinements were to spread any further in
Nebraska.

The 1997 study hearings had provided about a dozen senators on the Natural Resources and Agriculture Committees with stacks of documents reflecting multiple points of view on the scientific, economic, social, environmental, and public health issues connected to concentrated swine production. But a report on the study was never written, so only the senators who had attended the 1997 hearings had convenient access to the information. Others were left to find it on their own or with the help of staff and lobbyists.

One obvious source of scientific research on the issues was the Nebraska Department of Environmental Quality. NDEQ director Randolph Wood was a licensed professional engineer (the state registry of professional engineers shows his expertise was in aeronautics). A column that Wood wrote for the *Omaha World-Herald* in the fall of 1997 may have reassured many Nebraskans, no matter what their position on big hog farms. He wrote that the NDEQ was doing a good job of protecting water from pollution and that livestock waste lagoons, if properly constructed, "effectively treat biological wastes without causing an adverse effect on the environment."[1] He said a "rigorous regulatory program" had helped Nebraska avoid the problems experienced by other states.

But in early January 1998 Wood offered a dramatically different assessment in the report the governor had ordered a month earlier. Acknowledging that his conclusions wouldn't be popular with livestock interests, Wood told Governor Nelson that scientific studies from Nebraska, Iowa, and Minnesota concluded that livestock waste lagoons leak.[2] Wood said Nebraska's quarter-inch-per-day limit on lagoon seepage translated to "6,800 gallons of leakage per day for a one-acre structure or nearly 2.5 million gallons per year."[3]

He wrote that the risk of groundwater contamination increased with the size of a lagoon and when lagoons were built on sandy soils or above a high water table. Wood said the NDEQ should require monitoring wells around hog waste lagoons if their location and size increased risk to the groundwater. Because the state already required monitoring wells around landfills and municipal waste lagoons, Wood said it was logical to require them for some livestock operations. He would later tell senators that, in twenty-five years of having authority to do so, the NDEQ had never required a monitoring well for a livestock waste lagoon.[4]

Wood was also concerned about threats to surface water. Most of Nebraska's approximately five hundred lakes suffered from some degree of eutrophication, that is, an excess of nitrogen and phosphorous (called "nutrients") which feed algal blooms.[5] When the algae die, their decomposition decreases oxygen in the water, killing fish and other aquatic life. The EPA says nutrient runoff from livestock operations is a primary cause of eutrophication of lakes, rivers, and streams nationwide.

The Clean Water Act gives states the authority to regulate runoff (so-called nonpoint source pollution), but many states have been slow to do so until environmental groups have sued to force compliance. Although nitrogen contributed to the problem in Nebraska, Wood said excess phosphorous from nonpoint source pollution was "the primary concern." He said the NDEQ should require livestock operations to limit the phosphorous they put on cropground to what growing plants can use.[6]

Opponents of industrial hog farms who knew about Wood's findings welcomed them. But livestock groups complained about the cost of limiting phosphorous, digging monitoring wells, and providing plans for emptying waste pits when operations closed. As for Wood's assertion that risk increased in proportion to size, the livestock lobby argued that big operators tended to be less damaging to the environment because they could afford to hire professional engineers and managers to design and run their facilities, using the latest technology.

Although Governor Nelson had said he wanted to reassure the public by producing an objective statement of scientific facts, Wood's report wasn't widely distributed. Transcripts of hearings and legislative debate show little evidence that members of the legislature considered or even knew about the recommendations of the state's chief environmental officer as they debated how to regulate a growing mountain of manure. Many senators pledged to rely on "sound science" to guide their lawmaking on the topic, but the public record reveals little about how they determined what science to rely on as the hog-farm controversy began to dominate their agenda.

In interviews, senators admitted they were sometimes confused by the conflicting scientific points of view; they used various methods for choosing what to believe.

Omaha senator Jon Bruning relied upon what lobbyists for the Farm Bureau, cattlemen, and pork producers told him. "I felt like I was getting honest information from them," said Bruning.[7] Although Senator Ed Schrock said he didn't know whether phosphorous was a problem in the state's surface waters, he said he relied upon information from the NDEQ.[8]

Jerry Schmitt of Ord, who liked to hunt and fish, was concerned most about protecting water quality. He received his information from acquaintances who belonged to Mid-Nebraska PRIDE. "I guess I was listening to the farmers out here and kind of took their advice over what I was getting from the lobby and the administration of these farm groups," said Schmitt. "Common sense, I think, played a big part in this, too."[9]

Other senators said they relied on their own experiences either as farmers or as relatives of farmers when determining whose scientific information to believe. This method resulted in senators taking a range of positions on the question of regulation.

For example, Senator Jim Jones, who had a two-thousand-hog confinement operation of his own near Eddyville, said that if the phosphorous levels in his soil started to rise from too much hog manure, he would plant alfalfa to take up the excess nutrient. Although he agreed that most farmers probably wouldn't monitor phosphorous levels unless required by law, Jones came down on the side of less rather than more regulation.[10] He was among the thousands of Nebraska livestock producers who had never applied for a permit from the NDEQ until prompted to do so by the 1998 legislature.[11]

Hastings senator Ardyce Bohlke, a farmer who had initiated the legislature's 1997 study of hog confinements, said she read most of the material she received on the issue. One item particularly troubled her—Sand Livestock's practice of putting hog carcasses in deep concrete pits until they decomposed and then putting the liquid remains through center pivots with the rest of the hog waste. This method of carcass disposal wasn't included among those sanctioned by state law: burning, burying, or rendering. Bohlke thought of how her grandparents, who had raised hogs, would have guided her as she pondered the possible risks to human health posed by this practice.

"My grandmother used to have a phrase," said Bohlke. "If she thought I was being totally ridiculous, she'd say, 'Well, girl! Girl! Who wouldn't realize that that wouldn't be good?' I don't know eventually what all the scientific data will show," said Bohlke, "but you shouldn't put people at that kind of a risk, I don't think."[12]

In fact, state officials didn't know whether Sand's method of carcass disposal posed a risk to human health. Dr. Larry Williams, the state veterinarian, said Sand's method of disposal—not legal in any other state—probably had the same effect as burial. "Decomposition—whether it's in a pit or in the ground—goes through the same process," said Williams.[13] Although state law gave the Department of Agriculture the authority to regulate carcass disposal, Williams couldn't say whether any pathogens—such as salmonella or coliform—survived in the effluent from the dead pits.

For Omaha senator Don Preister, caution was the only logical approach when considering what science to accept. After sitting through all the public hearings and studying as much material as he could find on the topic, he said the legislature "applied science backwards." Preister said, "Frequently, I think the approach with agriculture is that [any practice is] OK until we prove it isn't."[14] He recalled that farmers had often overapplied nitrogen fertilizer before scientists discovered how easily nitrate can contaminate groundwater.

In the absence of conclusive research on the topics that concerned him most—lagoon safety and the use of hog waste on cropground—Natural Resources Committee chairman Chris Beutler said it made sense to put a process in place so the NDEQ could adjust its rules as conclusive research became available. But the livestock industry was hostile to any new regulations. "Just to get to something that was within the realm of common sense was a politically difficult thing to do," said Beutler. "It was like, who cares about good science? We're not close to it, anyway."[15]

Although policy makers typically seek firm answers to questions about how to protect the environment and human health, scientists often disagree —whether the issue is global warming, causes of cancer, automobile safety, or any of the dozens more scientific issues that contemporary policy makers deal with. In 1995 experts in occupational and environmental health, economics, anthropology, sociology, biology, agricultural law, and engineering from a dozen universities in six states, including Nebraska, had called for more research to help both scientists and the public understand the impacts of large-scale swine production.[16]

But the wheels of research move slowly. In early 1998 scientists hadn't yet conclusively answered many important questions about the safety—both to the environment and public health—of many waste management practices commonly used in hog confinements. Answers were just beginning to emerge from publicly funded research underway in Iowa and North Carolina. The only thing that experts seemed to agree on was that pig manure stinks, and that methods of limiting odor still needed to be perfected.

In the end, science took second place to economic considerations in the 1998 Nebraska legislature's debate over what Senator Bohlke called the "pig poop" issue. At a time when the so-called state-of-the-art technology for handling manure at big hog farms was losing favor in states with a record of burst lagoons, waste spills, dead fish, contaminated groundwater, and polluted streams, that same technology would spread across Nebraska with the blessing of the legislature.

The focus on economic issues would be managed by a coalition of some of the most powerful business interests in the state. On a cold morning in late January 1998, a press conference on the west side of the capitol brought together the Nebraska Bankers Association; the Nebraska Chamber of Commerce and Industry; the Omaha Chamber of Commerce; the Farm Bureau; the Nebraska Cattlemen; the Nebraska Pork Producers Association; corn, soybean, and sorghum growers; and dealers in farm equipment and fertilizer.

The group called itself "CLEAN" (Coalition for Livestock, the Environment, and Agriculture in Nebraska). CLEAN announced it supported a bill that would essentially change nothing in the way the state regulated livestock waste. Appearing with the group was Governor Nelson's chief spokesman on farm policy—Agriculture Director Larry Sitzman. Gesturing happily at the crowd around him, Sitzman said agricultural interests had "never been as united as they are this morning. . . . These people represent thousands of Nebraska agriculturalists."[17]

CLEAN favored a bill sponsored by Senator Stan Schellpeper, an amiable farmer, cattle feeder, and former president of the Stanton County Livestock Feeders Association. Schellpeper, along with twenty-four senators who signed onto his bill and livestock-industry representatives who helped write it, wanted to make minor adjustments in the current law and to provide two hundred thousand dollars for the NDEQ to hire more people to enforce the law. The bill would have authorized the governor to appoint livestock producers, scientists, and other citizens to advise the NDEQ on how to regulate the industry. Schellpeper also proposed a ninety-day deadline for the NDEQ to issue permits—a measure supported by hog producers eager to build new confinement farms or to expand old ones. Schellpeper's bill did little to satisfy "agriculturalists" who hadn't been invited to join CLEAN, including the Center for Rural Affairs, the Nebraska Farmers Union, and Mid-Nebraska PRIDE.

Those groups were attracted to two other bills. One sponsored by Lincoln senator Chris Beutler would for the first time require livestock producers to pay the NDEQ for inspections and permits. Bigger operations would pay higher fees so the NDEQ could hire enough staff to do the work. For the Environmental Quality Council—the citizen group appointed by the governor to adopt NDEQ regulations—Senator Beutler proposed several tasks. He wanted them to set rules for applying livestock waste to land, reducing runoff and odor, monitoring groundwater, and assuring that big livestock operations would pay to clean up any spills. In another bill, Senator Bohlke also proposed fees, saying it was only fair that the livestock industry should pay for regulation just like many other state-regulated industries such as banking, barbering, medicine, real estate, and insurance.

A twenty-five-year-old state law required all livestock producers to ask the NDEQ for an inspection to determine if they needed a facility to contain manure to prevent it from polluting surface water and groundwater. By now it was clear that thousands had ignored the law. Bohlke proposed giving them "amnesty"—several months to come into compliance without penalties.

Bohlke, Beutler, and Schellpeper all wanted to deny permits to "bad actors," that is, operators who habitually or intentionally violated environmental laws in Nebraska or elsewhere. All three bills were called the "Livestock Waste Management Act."

On 30 January, at the Natural Resources Committee hearing on the three bills, public comments repeated most of what had been heard during the 1997 study. The capitol hearing room overflowed with opponents and supporters of big hog farms. Fears for the environment, the quality of rural life, human health, and the future of family farms were countered by assertions that the concentrated feeding of pigs posed little environmental risk and that unnecessary restrictions on livestock operations would drive them and the packing plants they fed to other states. The Nebraska Cattlemen opposed further regulation of hog farms; they knew any new requirements for swine would probably affect cattle operations as well. Cattle feeders said that because environmental laws benefited the entire state, it was unfair to expect livestock producers to bear the financial burden for complying with them.[18]

More evidence emerged that the state's twenty-five-year-old livestock waste law had been only randomly enforced. Randolph Wood said that until 1997 the NDEQ had issued about fifteen hundred permits. The agency issued an additional 158 permits in 1997 but did no inspections to be sure people were complying with the law, except to respond to complaints from the public—"an absolutely unacceptable situation."[19] Wood said current law required the NDEQ to regulate all nutrients, including phosphorous, but it had regulated only nitrogen. He said the agency's failure to adequately enforce the law was due to "historic inertia" and to a lack of adequate staffing.

Absent from the hearing were pork producers whose expansion had generated the controversy—Chuck Sand, Brian Mogenson, Rich Bell, Jim Pillen, and Mike Bowman. Most of them would make their opinions known through hired lobbyists when a compromise bill reached the full legislature.

The Natural Resources Committee took two months to write Legislative Bill 1209, a version of the Livestock Waste Management Act that included elements of all three bills that had originally gone by that name. LB1209 included fees ranging from three hundred to five thousand dollars, based on the size of the operation; classification by the number and type of livestock (for example, sows, nursery pigs, feeder cattle, or turkeys); a mandate to regulate the biggest operations first; the amnesty that Bohlke had proposed; more staff for the NDEQ; a committee to advise the NDEQ on industry practices; a mandate for the agency to develop regulations for odor and land-application of waste; a bad-actor clause; and a requirement (called "financial assurance") that livestock producers keep enough money in reserve to pay to clean up a facility if they quit raising livestock.

Only five of the eight committee members voted to move the bill to the legislature for debate, reflecting a split in opinion that would carry over to the full legislature.

Bryce Neidig, the president of the Nebraska Farm Bureau, attacked the bill in a guest editorial in the *Lincoln Journal Star*, saying the proposal was misguided "regulatory overkill" because it imposed new fees and restrictions on farmers. He said the best way to protect the environment from livestock waste would be "through voluntary, incentive-based methods." He said livestock producers were "under siege by federal, state, and local regulatory proposals" and that the bill threatened the economic well-being of the entire state. [20]

On 30 March, when the legislature began debate on LB1209, the capitol rotunda was crowded with people who had engaged in the battle to that point. In a season when they would typically be preparing to plant crops, farmer-members of Mid-Nebraska PRIDE came to Lincoln as unpaid citizen lobbyists. Having lost their plea for emergency zoning powers with the death of LB1152, they asked for more staff for the NDEQ, fees to pay for enforcement, local public hearings on permit requests, a requirement that hog producers have NDEQ permits in hand before building, financial assurance, and restrictions on odor. They rubbed shoulders with pork and cattle producers who feared that rash action by the legislature would hurt their opportunities and income.

Professional lobbyists representing the state chamber of commerce, the livestock organizations, the Farm Bureau, corngrowers, and the biggest hog producers were also there. Premium Farms and three of the nation's biggest pork producers—Sand Livestock, Bell Farms, and Progressive Swine Technologies—spent nearly $112,000 lobbying the legislature in 1998. Lobbyists hired by the Nebraska Cattlemen and the Nebraska Pork Producers Association raised the total that was spent lobbying on livestock issues to

$134,862.[21] This is a small fraction of the $6.8 million that special interests spent on lobbying the 1998 legislature, but it was the first time anyone recalled seeing this level of activity from livestock interests.

It is clear which group had the attention of the majority of senators. Several times during a debate that lasted almost two days, senators commented on what livestock producers would or wouldn't support and what changes they wanted in the bill. Senator Ardyce Bohlke, who had come through the study and the LB1209 hearings with her concerns intact, called LB1209 "the industry's bill."[22] Under intense pressure from the livestock lobby, Senators Schellpeper and Beutler led efforts to negotiate a compromise, hastily drafting amendments as they were suggested by the lobby.

The prevailing argument was the need to protect the state's six-billion-dollar livestock industry from costly regulation. About half the farmers in the legislature joined that side of the argument while the other half supported more restrictions. For example, Senator Cap Dierks of Ewing, a rancher and veterinarian, wanted more stringent environmental controls; he was being flooded with complaints from his northeast Nebraska constituents who said Premium Farms was building without permits. On the other hand, no one was building hog barns in Senator Roger Wehrbein's district in the southeast. Wehrbein said, "I think it gave me a freer ability to try to do what's right, what's sustainable, what is cost effective, what's somewhere in the middle."[23] Wehrbein, whose farm near Plattsmouth included a small dairy herd and some hogs, advocated accommodations to help the livestock industry thrive.

Although the opposing factions in the rotunda differed on how the state should regulate livestock waste, they seemed to agree on one goal: the survival of small- and medium-sized farms. The Nebraska Farm Bureau, the Cattlemen, and the Pork Producers said stringent environmental laws entailed expenses small operators couldn't afford. But these groups also argued that to be fair the state should apply the same regulations to all farms, regardless of size.

In contrast, the Farmers Union, the Center for Rural Affairs, and PRIDE wanted to protect small farms by exempting them from regulation. They advocated a view of thriving rural communities based on environmentally friendly small- and medium-sized family farms. With his emergency zoning bill dead, Senator Jerry Schmitt continued to press for more rules for bigger operations. "I don't think fifty thousand head is a family farm," he said. "I think it's a factory and I think it should be regulated as a factory."[24]

But other senators argued that the trend toward bigger farms was inevitable because they were more efficient and profitable; they said it was inappropriate to adopt environmental policy as a substitute for economic policy. After two days of extended debate and intense efforts to write a bill that the livestock industry would accept, LB1209, much amended, was passed on a vote of thirty-nine to six.

The bill included money for twelve new inspectors for the NDEQ and set time limits for the agency to process permit applications so that producers who wanted to expand wouldn't be delayed. LB1209 established four sizes of livestock operations, with inspection and permit fees ranging from fifty dollars to five thousand dollars, depending on the number of animals in the operation. The bill also allowed existing unpermitted operations to avoid fees and penalties by contacting the NDEQ for an inspection by 1 January 1999. LB1209 required the NDEQ to notify the county board and the local natural resources district when a livestock producer asked for a construction permit to manage the waste of more than 5,000 animal units (12,500 pigs).

LB1209 gave the NDEQ authority to deny permits to applicants with bad environmental records—the so-called bad actor clause. A permit—once granted—would be good for as long as the operation existed unless the NDEQ had "cause" to revoke it.

The law allowed effluent composed of liquefied hog carcasses to be injected below the surface of the ground but barred spreading it on the surface. With this provision, introduced by Senator Bohlke, Nebraska became the only state to legalize carcass liquefaction.

The legislature also retained the quarter-inch-per-day seepage limitation for waste lagoons, equal to that in force in Kansas and one of the highest in the nation. Other states' seepage rates ranged from one-fifty-sixth inch per day in Minnesota to one-thirty-second inch per day in Colorado and one-sixteenth inch per day in Iowa.[25] To what degree the seepage carried pollutants to the groundwater was still not understood.

The 1998 Nebraska legislature rejected setbacks from residences and drinking wells (which would have put some distance between neighbors and pigs). It also rejected local public hearings on permit requests and financial assurance for cleaning up livestock operations that closed. All had been requested by the grassroots activists. Pork and cattle producers successfully argued that such mandates would unduly restrict free enterprise and would be too costly.

There's no evidence in the public record that senators were aware of the results of a study they had requested: the 1996 Nebraska Rural Poll found that over half of rural Nebraskans (58 percent) didn't believe that environmental regulations should be relaxed to reduce compliance costs for business. The poll also found that more than 40 percent of rural Nebraskans thought the state wasn't doing enough to protect groundwater from pollution.[26]

LB1209 required producers to use "best management practices" to reduce odor—but only until 1 July 1999. The director of the Nebraska Pork Producers Association complained about the requirement. He wrote that there wasn't enough good science on odor control to support giving the NDEQ any authority, and that "This would be similar to giving the Nebraska State Patrol authority to enforce speed on the highways, even if there was no such thing as the speedometer."[27]

With LB1209 the legislature set up a task force to make recommendations on controversies that senators hadn't fully resolved: the fees, if any, that livestock producers should pay for permitting and inspection; how to control odor; how to pay for cleaning up manure spills and abandoned lagoons; and how best to dispose of dead animals. The nine-member task force, made up of livestock producers, university scientists, local government officials, and an environmentalist, would have nine months to conduct its study and report to the legislature.

The legislature would take up the hog-farm issue again in 1999. By then the controversy would have spread across most of the state. NDEQ would grant another 142 permits for new livestock operations in 1998. Of the eighty-six applications for hog confinements, fifty-seven proposed housing more than a thousand head.[28] Bowman Family Farms of Colorado would eventually abandon its proposal to build barns for more than half a million hogs in Dundy County. But Bell Family Farms proceeded with plans to house eighty-four thousand hogs in Chase County.

Enterprise Partners—a company connected to Sand Livestock—proposed building a five-million-dollar farrowing operation in the Sandhills. The facility would accommodate fifty-seven hundred pigs producing 2.7 million gallons of waste on a hilltop in Arthur County where less than thirty feet of sand would separate the bottom of the waste lagoon from the Ogallala Aquifer.[29]

LB1209 did little to silence the controversy over what should be done to protect the state's air and water from pollution by hog manure. But the legislature's concern for the economics of the livestock industry was timely.

Hog production had been expanding nationwide for several months, providing more pigs than packing plants could handle. By late March 1998 the price of hogs had dropped to about thirty-one cents a pound—at least ten cents below the break-even point—and dozens of small hog farmers were leaving the business. By the end of the year, at prices as low as eight cents a pound, hogs would be less profitable than in the midst of the Great Depression. Many small hog farmers gave their pigs away rather than feeding or selling them. One farmer in Iowa proposed allowing the public onto his farm and charging them a fee to shoot hogs.

Expansion continued, however, among big confinement operators whose economies of scale allowed them to lose money for a while and remain in business.

It's a case of one sacred cow, which is the family farm,
locking horns with another sacred cow, which is the Nebraska
Constitution.
– Glenn LeDioyt, Chairman of the "No on Initiative 300"
Committee, quoted by Thomas A. Fogarty, *Lincoln Journal,*
8 September 1982

Hog Hiltons and Initiative 300

Brian Mogenson and his father, Harry, didn't set out to make history. They just wanted to make money raising pigs. In 1999 the Mogensons had state permits for more than 130,000 hogs at sites in four northeastern counties and were on their way to ranking among Nebraska's biggest pork producers.[1] But their plans stalled when Nebraska's attorney general said the Mogensons didn't qualify as a "family farm" under the state constitution.

In 1999 Brian and Harry Mogenson were the first farmers ever to be sued by the state for allegedly violating Nebraska's seventeen-year-old constitutional ban on farming by nonfamily corporations, commonly known as Initiative 300 or I-300. Nebraska Attorney General Don Stenberg said the Mogensons' limited liability companies—Nebraska Premium Pork and Premium Farms—would qualify as family operations only if one of the owners lived on the farm and provided day-to-day labor and management. Neither of the Mogensons lived on any of their ten hog farms. To comply with the law, the men reorganized their companies as general partnerships—a business organization legal under I-300 because it removes the liability shield that corporate business owners enjoy.[2] In a partnership, individual partners are liable for debts and environmental problems of the partnership.

Reorganizing his companies didn't end Brian Mogenson's troubles with the attorney general. A few months later the state began asking who owned the thousands of pigs in the barns at the Premium Farms sites. Neighbors had seen trucks from the Seaboard Corporation—the nation's third-largest pork producer—coming and going at Premium Farms. The neighbors gave their evidence to the attorney general. He concluded that Seaboard and Mogenson had gone to great lengths to conceal Seaboard's ownership of the pigs—illegal under Initiative 300.

In August 2001, rather than admitting to any wrongdoing, Seaboard and two subsidiaries signed agreements with the attorney general saying they would no longer do business with the Mogenson companies nor would they engage in farming or ranching or own agricultural land in Nebraska.[3]

In July 2001 the hogs that may or may not have belonged to Seaboard were already gone and Mogenson hoped his troubles with the attorney general were over. "I don't own any pigs," he said. "All I am is a Hog Hilton—I rent motel spaces out for pigs."[4] Mogenson, like any other Nebraska farmer, can contract to rent space for hogs, but those hogs can't legally be owned by a nonfamily corporation.

Brian Mogenson was an ambitious, hardworking entrepreneur and no fan of Initiative 300. He said it unfairly interfered with his own plans and ruined family farms by limiting farmers' ability to adopt new industrial models of agriculture. With his father, who died in 2000, Mogenson had built hog barns and feed mills in Iowa, which had few effective barriers to corporate farming. Mogenson said an Iowa farmer could build two hog sheds financed by a corporation like Farmland or Murphy.

"So the farmer owns the buildings and pays the taxes and Farmland puts in the bacon and takes the risk on price because the lease is locked in," said Mogenson. "In seven to ten years, the buildings are paid for and the hell with raising corn, because a farmer can say, 'I'm making eighty thousand dollars walking out my back door. A piece o' cake.'"[5] With each building paying thirty to forty thousand dollars a year, Mogenson said, it's a way for farmers to make a steady income and stay on the land.

By 1998 big swine farms like those run by Brian Mogenson had revived an older Nebraska controversy over who should farm the land. So intense was the public debate that the Nebraska Supreme Court once said, "the voters have been subjected to tornadolike winds in voting on this highly political question."[6] Although Nebraska's ban on nonfamily corporate farming had been law for sixteen years, the state did little to enforce it until pressured to do so by the public outcry over big hog confinements. To understand how that constitutional ban affected hog production in Nebraska, it is useful to review the history of Initiative 300.

In 1982 it would have been hard to find in Nebraska, or for that matter in the United States, an advocate for the demise of the family farm. For Nebraskans—as for many Americans—family farms represent democratic ideas that trace to the country's founding. The yeoman farmer, owning and working the land, was Thomas Jefferson's ideal citizen—hardworking, principled, thrifty, and inventive—a heroic figure, feeding the nation and the world.

Despite the popular support of family farming, by 1982 thousands of Nebraska's 47.5 million acres of farmland had been acquired by nonfamily corporations. Estimates of the corporate holdings ranged from 238,160 acres to 1.3 million acres.[7] By its own reckoning, the Prudential Insurance Company owned more than 33,000 acres of farmland in Nebraska and about 750,000 acres in other states.[8] By 1982 Prudential was one of several nonfamily corporations that had converted thousands of acres of Nebraska rangeland to cropground by installing center-pivot irrigation.

The trend alarmed some Nebraskans, who believed that the tax and liability benefits granted to corporations and their access to capital gave them undue competitive advantages over family farmers. There was an outcry among farmers and conservationists who believed that marginal land—especially in the Sandhills—would erode under heavy cultivation by a business entity concerned more about short-term profits than about conserving the land for future generations of farmers. In the *Polk Progress*, a small-town Nebraska paper with a populist leaning and a nationwide readership, editor Norris Alfred questioned "the wisdom of changing rangeland into cropland, particularly when the land is then planted to a crop already in surplus." He wrote, "The state and nation do not need more corn. . . . We would rather have an individual farmer raise a crop of hell on a small parcel of land than have a corporate 10,600-acre spread raise 161 bushels of corn per acre on land that is neat and lifeless."[9]

By the early 1980s eight states had statutory restrictions on corporate farming: Iowa, Kansas, Minnesota, Missouri, North and South Dakota, Oklahoma, and Wisconsin.[10] For about ten years various farm groups had tried unsuccessfully to persuade the Nebraska legislature to ban nonfamily corporations from farming, but Nebraska attorney general Paul Douglas said such a law would be unconstitutional. Neal Oxton, president of the Nebraska Farmers Union in 1981, recalled a conversation with Douglas. Oxton said, "I asked Mr. Douglas, 'How is it that eight states surrounding us can have laws against corporate farms and we can't here in Nebraska?' And he said, 'That's easy. They just have weaker constitutions than we do.' "[11]

At that moment, Oxton said he decided Nebraskans should amend their constitution to ban nonfamily corporate farming. Delegates to the Farmers Union's annual meeting in December 1981 enthusiastically approved Oxton's plan.

The Center for Rural Affairs loaned a staff member, Chuck Hassebrook, to help write what would be called the "Family Farm Preservation Act" and designated as Initiative 300 by the secretary of state. The language reflected the many bills that had failed in the legislature. Hassebrook said, "In essence, what Initiative 300 said was that anybody who wants can invest in and compete in agriculture, but . . . they're going to do it by the same rules that most family farmers do it—that is, without the legal liability protections of the corporation and by paying taxes as an individual, without the tax advantages of a corporation."[12] The drafting team included University of Nebraska law professor James Lake, who helped craft language that would withstand legal challenges.

The Family Farm Preservation Act defined "family" as people related to each other "within the fourth degree of kindred according to the rules of civil law, or their spouses." (Supporters typically defined "fourth degree of kindred" as first cousins or closer.) If a family corporation owned the land, the majority of the stock must be owned by members of the family, one of whom must either live on the farm or ranch or provide daily labor and management. More than a thousand words long, the act included fourteen exemptions to the ban on corporate farming. Exemptions included family corporations, nonprofit and tribal corporations, experimental farms, poultry farms, land leased by alfalfa processors for producing alfalfa, land operated for growing seed, and mineral rights on agricultural land. It dictated roles for the secretary of state and attorney general in enforcing the ban. It reserved for the public the right to sue to enforce the law if state officials failed to do so.

Secretary of State Allen Beermann approved the petition language in February 1982. The Committee to Preserve the Family Farm needed 49,242 signatures by 2 July to put the measure on the November ballot. They set a 25 June goal of 60,000 to be sure there were enough valid signatures.[13]

To organize the petition drive, the Nebraska Farmers Union hired Drey Samuelson, a young Nebraskan who had worked in the presidential campaigns of George McGovern and Jimmy Carter and as South Dakota congressman Tom Daschle's Sioux Falls coordinator. The early strategy was to send volunteers with petitions to wherever people congregated—

supermarkets, cafes, and sporting events, for example. But that method gathered signatures too slowly. So Samuelson devised a way to quickly expand the network of people carrying petitions.

He said, "Every time someone signed a petition who showed an ounce of enthusiasm for it at all, our people would ask, 'Would you take a petition?' We'd give them one right on the spot with the instruction sheet, and we'd put a star by their names so we knew who had petitions so we had a list of all of them. Then we started direct mailing and calling and urging them to get going, and update them."[14]

An estimated two thousand Nebraskans carried the petition.[15] The Farmers Union, the Center for Rural Affairs, the National Farmers Organization, the Grange, the American Agriculture Movement, several church groups, and Women in Farm Economics (WIFE) supported the effort. WIFE pulled its 1982 state convention out of Kearney because the chamber of commerce there opposed I-300.

The Nebraska Farm Bureau and livestock groups were officially neutral, although several of the state's biggest cattle feeders and ranchers opposed the measure. The many other opponents of I-300 included several state senators, bankers, real estate companies, insurance executives, and business groups. They said the detailed, specific language of I-300 belonged in statute rather than in the constitution. They said the exemptions alone would generate numerous lawsuits. Arguing that there was no logical reason to treat farming differently from other business ventures, they asked why family farms should receive protection not available to any other family-owned business, such as grocery stores, shoe-repair shops, or filling stations. Furthermore, critics said the proposed amendment violated the cherished American ideals of free enterprise and property rights.

But in an editorial that reached about ninety thousand Nebraska households, the *Catholic Voice* asked, "What possible meaning will free enterprise have if the trend of concentrating land ownership in fewer and fewer hands continues unchecked? Diversity in ownership and justice in the food production system are values that public policy should protect."[16] The petition drive received a boost when both candidates for governor—Democrat Bob Kerrey and the Republican incumbent Charles Thone—signed on.

It is important to point out that both critics and supporters of a ban on nonfamily corporate farming generally agreed upon the merits of family farms. They believed it was prudent to encourage policies enabling individuals to own and work the land, that many people doing so on many farms was preferable to massive tracts of land being farmed by tenants employed by corporations, that family farms had contributed much to the success of

American agriculture, and that family farms were good places to instill in children such qualities as industry, frugality, self-sufficiency, and neighborliness. In an editorial opposing I-300, the *Lincoln Journal* stated, "Farming and ranching by family units owning their own real estate is and should remain the preferred mode of agricultural production. Besides having several highly desirable social and cultural consequences, it's arguably also the most efficient farming and ranching system."[17] Opponents said I-300 created false hope among farmers; it would do nothing about high production costs and low prices for grain and livestock, which put family farms at much greater risk than corporate farming ever would. But the message "help us save the family farm" was simple and meant something to everyone.

By the end of May a little more than 30,000 signatures had been collected—only about half the goal. Drey Samuelson said, "I thought we were doomed." But then, "It was like the sky opened up over a dry field. Every day we got thousands of signatures in from people we didn't know personally."[18] By the end of June 62,543 people had signed;[19] 56,636 of the signatures were valid—more than enough to put the issue on the November ballot.[20]

The absence of organized opposition had helped the petition drive succeed. Then in September, two months before the election, a coalition of financial institutions, insurance companies, agribusinesses, prominent livestock producers, and others formed the "No on Initiative 300" Committee. To lead the campaign, the committee hired Glenn LeDioyt, president of an Omaha farm-and-ranch-management company and the finance chairman of the state Republican Party. He said they might spend as much as $500,000 to defeat I-300.[21] Insurance companies led the way, with Prudential giving $125,000 to the cause. Metropolitan Life and the Travelers Insurance Company each gave $35,000. By late September ads opposing I-300 swept across the state in newspapers and on radio and television. The No on Initiative 300 Committee bought so many ads that, under federal law requiring equal time, the Farmers Union received free air time to tell its side of the story.

All but a handful of the state's newspapers opposed I-300. They called the language of the proposed amendment "confusion and silliness," "cluttered," and "much too particular for inclusion in a constitution."[22] One newspaper predicted it would bring "economic disaster."[23] Despite the editorial opposition and the powerful ad campaign against I-300, an *Omaha World-Herald* poll in early October showed that two out of three Nebraskans favored it.[24]

In the three weeks before the election, Prudential kicked in another $145,000 for the anti–Initiative 300 campaign. Company officials said they were protecting their investments in Nebraska. In all, the No on Initiative 300 Committee spent a record-breaking $448,943.48 on the effort, with Prudential's contributions totaling $270,000. The Committee to Preserve the Family Farm would be outspent almost six to one.[25]

Nevertheless, in the election on 2 November, Initiative 300 won 56 percent of the vote. It won in seventy-two of the state's ninety-three counties, but lost in most of the Sandhills and panhandle, indicating a split in opinion between farmers and ranchers.[26] The cochairman of the No on Initiative 300 Committee, Wauneta cattle feeder and rancher Jack Maddux, said, "I am sorry that Nebraska citizens gave away people's rights. We'll never get them back again. They're gone forever."[27] But Edward Tvrdy, state president of the National Farmers Organization, said, "I think it was an issue of money versus people and the people won."[28] It was a sentiment echoed by Drey Samuelson, who said the election showed "the people defeating raw power."[29] The *Omaha World-Herald* said, "Reason was a casualty in the I-300 vote. Emotionalism prevailed."[30]

It was the eighth time Nebraskans had changed their constitution by the initiative process. Although unpaid volunteers had carried the petitions, there were some prizes. The Nebraska Farmers Union put the name of every petition circulator in a hat for a drawing of four twenty-five-dollar savings bonds. Every Farmers Union member whose name was drawn also received a paid trip to the state convention in Grand Island.

Initiative 300 officially became Article 12, Section 8, of the Nebraska Constitution on 29 November 1982. The victory didn't silence the opposition. Line by line, Initiative 300 was challenged in court. Each time it withstood the challenge.

In a 1986 lawsuit, Omaha National Bank argued, among other things, that the language of I-300 belonged not in the constitution but in statute, that I-300 conflicted with the National Bank Act by limiting the bank's ability to manage farmland held in trust, and that it violated the equal protection clause of the U.S. Constitution. The Nebraska Supreme Court rejected those arguments, upheld I-300, and for good measure, cited the lower court's decision in the case—a passage often quoted by I-300 supporters: "The ultimate source of power in any democratic form of government is the people. Our Nebraska Constitution is a document belonging to the people. Subject only to the supremacy clause of the United States Constitution, the people may put in their document what they will. Even to the shock and dismay of constitutional theoreticians, the people may add provisions dealing with 'non-fundamental' rights, as well as provisions bearing the most tenuous of relationships to the notion of what constitutes the basic framework of government. The people may add provisions which legal scholars might decry as legislative or statutory in nature. But the people may do it nonetheless."[31]

Nebraskans' power to change their constitution was upheld again in 1991 when the United States Court of Appeals, Eighth Circuit, ruled against MSM Farms, Inc. MSM argued that there was no legitimate state interest in banning nonfamily corporations from farming. Again, the court's conclusions are worth quoting: "The people of Nebraska have made a reasonable judgment that prohibiting non-family corporate farming serves the public interest in preserving an agriculture where families own and farm the land. It is not for the courts to second-guess the wisdom of this judgment."[32]

Since 1991 litigation over Initiative 300 has centered on the hog industry. In 1994 a group of unrelated farmers in Dawson County wanted to form a cooperative to operate a hog-farrowing facility. A district court said they would qualify as a nonprofit corporation under Initiative 300. The Nebraska Supreme Court disagreed in a 1997 opinion that clarified the definition of "nonprofit corporation" in Initiative 300.[33]

In 2000 the Nebraska Supreme Court interpreted the clause in 1-300 that requires at least one family member of an incorporated farm to be "residing on or actively engaged in the day to day labor and management of the farm or ranch." In *Hall v. Progress Pig*, a majority shareholder of an incorporated five-hundred-sow farm managed business strategy, financing, and payroll, met annually with a geneticist, and handled the marketing of the pigs. But he lived three miles from the farm and did none of the daily chores such as feeding pigs, cleaning pens, or administering medications. In what has since been called the "sweat test," the court said, "To be 'actively engaged' as understood by the common layperson would require that a person actually conduct such activities on a daily basis, not designate such activities to other individuals."[34] Ruling that the mental and business activities of a business don't qualify as day-to-day labor, the court wrote, "Labor would encompass the physical chores attendant to the farm, and management would encompass the mental and business activities of the operation." 1-300 requires family farmers to be engaged in both. Critics of the decision say this test wouldn't be met by many of today's ranchers and farmers who live in town and manage their business by phone and computer and with occasional trips to the ranch or farm.

For fifteen years, enforcement of 1-300 was haphazard at best. A few months after 1-300 passed, the legislature repealed a 1975 law that required corporations to report their farmland holdings to the secretary of state. Allen Beermann, who then held the office, supported the repeal as a way to prompt the legislature to find a way to enforce 1-300.[35] But there was no further legislative action to that effect until 1998, when, in response to citizen concerns about industrial hog farms, the legislature passed a law requiring all corporations involved in farming in Nebraska to explain how they met the requirements of 1-300. The new law, LB1193, also gave the attorney general subpoena power for 1-300 enforcement and money for one attorney to do the work. Governor Ben Nelson vetoed the money, but the legislature restored it.

In 1999 the enforcement job went to William "Russ" Barger, a young attorney who grew up on his family's farm in southwest Nebraska. He began by investigating numerous citizen complaints about suspected violations of Initiative 300. Between 1999 and 2001 Barger led investigations that eventually forced Premium Farms, the Seaboard Corporation, and Christensen Family Farms—a Minnesota corporation—to comply with the law. Like Premium Farms, Christensen—the nation's eighth-largest pork producer—agreed to operate in Nebraska as a general partnership. In addition, for five years the partnership must make annual reports of its Nebraska business and financial dealings to the Nebraska attorney general.[36]

In all three cases, citizen complaints led to the attorney general's investigation. In fact, Barger says citizens are the primary source of information on possible violations of Initiative 300. It happens when "a citizen calls up and asks a question," says Barger. "Eight out of ten times, they know what they're talking about when asking us to look into it."[37] Barger tries to find objective, written evidence before investigating.

Initiative 300 assigned enforcement responsibilities to both the attorney general and the secretary of state. The corporate farming reports required by the 1998 law are stored with the secretary of state. But lacking the staff necessary to closely scrutinize the reports, the secretary of state has typically acted merely as a kind of file clerk for them.[38] Before the 1998 reporting law passed, the secretary of state annually referred one or two companies to the attorney general for investigation. Since the law passed, that number has risen to about a dozen a year.[39]

Volunteers from Friends of the Constitution—a coalition of about twenty agricultural groups and churches that advocate for the enforcement and continued support of Initiative 300—spend several days every year examining the reports. They have found many additional questionable farming and ranching activities and have pointed them out to the attorney general.

There are other barriers to 1-300 enforcement. Russ Barger is the only attorney assigned to the task of investigating highly complicated business relationships and transactions. Typically, those being investigated are reluctant to cooperate. Furthermore, the legislature hasn't established any penalties for violating the law. If a company is in violation, all it must do to satisfy the state is to change its organization or sell any land or livestock it owns. The secretary of state has the authority to dissolve any business that fails to comply, but has never done so.

Another barrier to enforcement involves the difficulty of tracing livestock ownership. Typically, livestock trades hands much faster than land, and there is no depository for documents related to those transactions. Documentation of landownership can typically be found in a county courthouse.

Despite increased efforts to enforce 1-300 since 1998, some want the state to do more. They wonder how farms with tens of thousands of animals can be legal under 1-300. But 1-300 was never intended to limit the size of farms, only their business structure and ownership. For example, Sand Livestock Systems, Inc., a corporation headquartered in Columbus, is among the twenty largest hog producers in the country.[40] The company has built hog operations in about a dozen Nebraska counties. A related company, Sand Systems, Inc., manages hog farms. Each farm is owned by a general partnership whose partners typically include some officers of Sand Livestock. The apparent close relationship between the partnerships and the corporation has raised some eyebrows, but no violations of 1-300 have been shown.

Other large nonfamily, corporate hog operations in Nebraska existed before 1-300 and were grandfathered in, such as Hastings Pork in Adams County and National Farms in Holt County. In 2000 National Farms was bought by Christensen Family Farms, Inc. It was at that point that Christensen reorganized its Nebraska business into a partnership to satisfy the requirements of Initiative 300.

Because 1-300 places no restrictions on the size of a farm or ranch, it does nothing to impede legitimate family hog-producers who want to raise thousands of their own hogs.

Nebraska's constitutional ban on nonfamily corporate farming is considered the most stringent in the nation, but Nebraskans still debate whether 1-300 did what it was designed to do—save family farms. Supporters say 1-300 has helped family farmers stay in business by making everyone play by the same rules. But critics say it has forced many farmers off the land by limiting their access to corporate capital and packer feeding contracts. A study of nine midwestern states with anticorporate farming laws concluded such laws are effective "if effectiveness is defined as limiting the acreage under non-family corporate ownership arrangements." But the study also concluded that there are grounds for people to disagree over the appropriate role for nonfamily corporations in agriculture. Thus, "state governments and citizens have clear policy options."[41]

Using 1-300 definitions, it is nearly impossible to tell how many family farms are in Nebraska. But because the state has a long tradition of family farming, it is useful to consider the statistics on the numbers of Nebraska farms. In 1982 there were sixty-three thousand farms in Nebraska; in 2000 there were fifty-four thousand farms—a loss of 14 percent. In Nebraska's bordering states, the decline in the number of farms during that same time ranges from 7 percent in Missouri to 23 percent in Minnesota.[42] (The number of farms in Wyoming and Colorado actually increased, but there was a decline in total farm acreage.) The overall loss in the number of farms and the corresponding increase in the size of farms has occurred despite the presence of anticorporate farming laws.

In 1998 South Dakota passed a constitutional amendment to restrict corporate farming that is patterned after Nebraska's. The Missouri legislature relaxed its law in 1993 to allow Premium Standard Farms to operate in the northern part of the state.

Polls have shown that the majority of Nebraska farmers continue to support Initiative 300 and that rural people support policies that help to sustain family farms.[43] But within all agricultural groups there are strong differences of opinion about Initiative 300. Even after twenty years it generates heated debate. The Nebraska Pork Producers Association officially supports Initiative 300, but individual members hold a variety of opinions about it.[44] Some other powerful groups representing agriculture and business continue to object openly to elements of Initiative 300. The Nebraska Farm Bureau and the Nebraska Cattlemen both favor amending it to conform to their philosophies supporting "the free enterprise system." The two groups believe 1-300 should be amended to make more capital available to people who want to begin farming and ranching and to enhance "value-added projects with producers in Nebraska." The Nebraska Cattlemen want 1-300 modified to enable the transfer of "assets from one generation to the next."[45] In its policy statement, the Farm Bureau says, "As long as there is a corporate farming amendment, we request that it be enforced."

Some bankers continue to object to constitutional limits on their use of farmland held in trust and see 1-300 as a barrier to free enterprise, but the director of the Nebraska Bankers Association said the issue is "too emotionally charged" for the group to take a position on 1-300.[46] Two bills to amend 1-300 have died in the legislature because the attorney general has said that any amendments to Article 12, Section 8, of the Constitution would have to be approved by voters—just as those who wrote 1-300 intended.

Norma Hall, who farms with her husband near Elmwood, has from the beginning supported the ban on nonfamily corporate farming. She carried a petition in 1982. Hall believes Initiative 300 is always at risk, in part "because we have fewer family farmers out there now than we did in those years."[47] Hall filed the lawsuit in the Progress Pig case that led to the Nebraska Supreme Court upholding citizens' right to sue and that defined day-to-day labor and management. She also belongs to Friends of the Constitution; with other volunteers, Hall has examined corporate reporting forms for possible violations of Initiative 300.

When Initiative 300 was passed in 1982, the arguments for and against it were based largely on the economics of farming, but concerns about the effects of turning Sandhills rangeland into cropground added an environmental component to the debate. It is possible, in fact, that Initiative 300 has helped Nebraska to avoid serious pollution from livestock waste as experienced by other states with less restrictive corporate farming laws. Chuck Sand—a long-time critic of 1-300—said, "It's bad for the state in general," but he also said, "We don't have the (sewage) lagoon spills and other things that make it bad for everybody."[48]

Opinions differ on that point, even among those who support Initiative 300. Chuck Hassebrook, the current director of the Center for Rural Affairs and one of 1-300's authors, said, "There's no question but that the presence of Initiative 300 dramatically reduced the extent of industrial livestock production in Nebraska, reduced the environmental problems associated with it, and reduced the conflict over it compared to states like Iowa."[49]

Marty Strange—the former director of the Center for Rural Affairs— said, "I wouldn't substitute 1-300 for effective environmental regulations." But he added, "I think 1-300 is part of an attitude in Nebraska that big agriculture ought to be distrusted. And I think that signal gets through in a lot of ways, and it creates an environment in which those kinds of operations don't or can't come to the state—or if they find a way to come to the state and comply with 1-300 they're awful careful."[50]

Many of those who support Initiative 300 encourage that caution by invoking the words carved in stone over the north entrance to the State Capitol: "The salvation of the state is watchfulness in the citizen."

This is historic litigation of epic dimension; it is a part of the evolving history of the Nebraska prairie. Just as barbed wire changed the face of the Great Plains, confinement livestock production [is] revolutionizing agriculture. . . . Yes, National Farms is big. But more hogs does not necessarily mean more odor.

– Ray Goeke et al. v. National Farms, Inc., Appellant Brief to the Nebraska Supreme Court, 7 July 1992

A Tale of Two Counties

Nebraskans who in the late 1990s were speculating about the short- and long-term effects of putting thousands of hogs in a limited space could have considered evidence from two counties with long experience of industrial hog farms—Holt County in the northeast and Furnas County in the south.

HOLT COUNTY

You can't be long in O'Neill before a cattle truck rumbles through the middle of town, stirring up dust on Highway 20 and leaving behind the rank odor of cow manure. Among all counties in the nation, Holt County ranks second in the number of beef cows and in the top fifty in cattle production.[1]

Cattle have not caused much controversy in Holt County—not since the range wars of the late nineteenth and early twentieth centuries. Today, cattle on pasture dominate the landscape of southern Holt County, where in numerous wet meadows the Ogallala Aquifer is so close to the surface that ranchers can simply drive a pipe into the ground to have an abundant supply of pure, cool water bubbling up for their cattle.

There is ample water in northern Holt County as well, but its purity has been compromised by three decades of intensive corn production, heavy use of nitrogen fertilizer, and center-pivot irrigation. By 1996 nitrate in the water beneath 102,000 acres of northern Holt County averaged 20.8 parts per million.[2] Ten parts per million is the federal safe-drinking-water standard for nitrate.

Nitrate pollution of the groundwater in northern Holt County is one result of the interconnection of corn, hogs, and irrigation that drives the economy there. Holt County leads the state in the number of acres irrigated through center pivots.[3] Among all counties in the nation, it's in the top fifty both in corn production and in the number of hogs.[4] All but a few thousand of the county's two hundred thousand hogs are housed in 165 buildings belonging to National Farms on two sites near Atkinson.

In 1982 National Farms, headquartered in Kansas City, came to Holt County with heavy financing from a bank in the Netherlands. While the facilities were being built, nearly 300 construction workers boosted the income of motels and restaurants in the area. National Farms promised to employ 150 people and to spend thirty-one million dollars a year on wages, feed, utilities, taxes, insurance, chemicals, breeding stock, and supplies.[5]

Business people in Atkinson and O'Neill championed the arrival of National Farms. Farmers could count on a new local market for about four and a half million bushels of corn each year.[6] There would be business for the hardware stores, filling stations, trucking companies, and sellers of irrigation equipment.

National Farms would eventually provide a $3.75 million payroll and benefits package to about 175 employees.[7] By 1986 the company farmed 131 quarters (20,960 acres) of Holt County cropground and had eighteen thousand sows in a farrow-to-finish operation that produced more than three hundred thousand market hogs a year.[8] Because its Nebraska operations preceded the passage of Initiative 300, National Farms operated legally as a corporation grandfathered in under the law.

Some Holt County residents feared that odor from the waste of so many hogs would make the air unbreathable; National Farms officials said they would build waste lagoons big enough to control odor and would situate them no closer than one-half mile from the nearest neighbors. In response to fears of groundwater contamination, a National Farms official said the lagoons wouldn't leak. He said it was in the company's own interest to prevent nitrate contamination of the groundwater because the hogs would have to drink it.[9]

The Holt County location matched all the criteria that National Farms chief executive officer Bill Haw used in siting a very large hog farm: abundant cheap feed, enough suitable land for disposing of the waste, nearby markets, a good-quality labor supply, and low population density. In fact, Haw's primary consideration was that the area should be thinly populated in order to lessen complaints about odor and flies and to avoid "the absolute unpredictability of what regulations, regulators, and the courts may decide to be an unsuitable nuisance."[10] In Holt County there were fewer than six people per square mile, but sparse population didn't spare National Farms from conflict.

Early complaints came to National Farms in a letter from the Holt County attorney. He threatened legal action because the company had allegedly laid plastic pipe in the county's right-of-way and had laid culverts and done dirt work on county roads without the approval of the county board. The county attorney also warned the company that it was putting other farmers' swine herds at risk of disease by improperly leaving dead hogs piled outside of barns and by failing to cover rendering trucks carrying carcasses away. The county board worried that National Farms' earthen lagoons would leak into the groundwater. The board also accused the company of spraying contaminated water onto tiled fields that drained into surface water.[11] The Holt County attorney sent a copy of the letter to the Nebraska Department of Environmental Control. (The name was changed to the Nebraska Department of Environmental Quality in 1991.)

The NDEC replied that the company's facilities met or exceeded the state's design standards and that the agency had no authority to deal with carcass disposal and odor, which is not to say that the agency was unaware of the problems.[12] An NDEC inspector visited the operation and wrote, "The odor wasn't too bad till you got down wind. I wouldn't say it would knock your socks off but it sure got my attention. . . . Problems such as dead hogs and odors are going to be continual due to the nature of the hog business and particularly with the size of these operations."[13] Because the state did nothing about the problems, they were left up to local authorities to resolve.

The Holt County district court became the arbiter. Between 1985 and 1992 twenty people representing ten families filed six nuisance lawsuits against National Farms. The litigating neighbors lived up to four miles from the hog farms—far beyond the half-mile range that the National Farms official had implied would be far enough away to avoid odor problems.[14] The neighbors said the odor made them sick. They said that odor, dust, and flies from the hog farms interfered with their enjoyment of their own property.

District Judge Edward Hannon refused to allow National Farms to introduce evidence of the company's economic effect on the community. During one trial, more than fifty local businessmen and National Farms employees gathered to protest in the snow outside the Holt County courthouse. They said National Farms' annual ten-million-dollar impact on the local economy deserved the court's consideration.[15]

Over the years, juries awarded more than eight hundred thousand dollars to the plaintiffs. National Farms was forced to spend millions to reduce odor and to settle some cases out of court. In only one of the lawsuits did a jury side with National Farms. The company appealed three decisions to the Nebraska Supreme Court, which each time ruled against National Farms and for its neighbors. Under court order to reduce odor, the company changed the nozzles on its center pivots to sprinkle waste down onto the crop rather than spraying it high into the air. They expanded their crop acreage so the waste would be less concentrated and changed the way the manure was handled. "Some years we did make the money back and some years we didn't," said former National Farms manager Greg Gilsdorf. "It was all a drain on the bottom line."[16]

There is plenty of anecdotal evidence that National Farms provided jobs for people who might otherwise have left the area and that some businesses profited when the company bought supplies or services from them. To draw attention to its economic impact on the community, National Farms once paid sixty-seven thousand dollars in payroll in two-dollar bills. As people noticed those bills circulating, the point was made. One local farmer said, "A lot of people cuss National Farms, but you shouldn't bite the hand that feeds you."[17] By many accounts the company was a good citizen, donating food baskets and hams at Christmas and money for a Little League dugout.

But a 2000 report by University of Nebraska sociologists came to mixed conclusions about National Farms' effects on the county. Holt County was one of thirty-six studied in six states over fifteen years. Among measures of prosperity, the study found that retail sales and the number of retail establishments were unaffected by increased hog production. But the researchers couldn't determine whether farm jobs increased. They found that population levels declined and property taxes rose, but the study also determined that per capita incomes rose faster and the percentage of people in poverty fell when the counties in the study were compared to counties with no growth in hog production.[18]

Nebraskans who supported increased pork production praised the study and said all doubters should be satisfied. But a Colorado economist said the study was so flawed "that no real conclusions can be reached."[19] In fact, John Allen, the lead author in the Nebraska study, said later, "The real finding of our study was that given the limitations of data it is almost impossible to sort out impacts of swine or other dramatic changes in agriculture on communities."[20] Allen also urged policy makers to be cautious in using the study because it didn't address how big hog operations affect the quality of people's lives. He wrote, "Concerns over odor and environmental quality, access to markets, and impacts on community harmony are legitimate" and more research needs to be done on those concerns, which economists call "externalities."[21]

One such externality is environmental impact. NDEQ documents show that a high level of groundwater contamination was detected beneath National Farms in the late 1990s when the company began testing groundwater from existing wells in the area. In an internal memo, an NDEQ employee wrote that the monitoring results "indicate large amounts of contamination possibly penetrating to great depths within the aquifer."[22] Levels of nitrate contamination as high as 120 parts per million were discovered in irrigation wells on cropground about a mile from the operation. But those wells and others nearer to the lagoons drew from too deep within the aquifer to precisely determine the source of the pollution.

In 2000 National Farms sold its Holt County operations to Christensen Family Farms. In 2001—two years after the contamination was discovered— the NDEQ required Christensen to dig new monitoring wells and continue to sample the water.[23] The agency will need at least two years of data before it can determine what caused the high nitrate levels around the operation.

Two possible sources of the contamination are commercial nitrogen fertilizer and animal waste, both of which can easily permeate the sandy soils of northern Holt County. Isotope tests, which can determine the source of the nitrate, haven't been used on the samples. NDEQ officials say the agency doesn't have enough funding to do the expensive tests, but it could require Christensen to do them. In situations like this, once the source of contamination is determined, the NDEQ could order a cleanup, but that possibility is years away.[24]

Whether the heavy nitrate pollution of the aquifer around National Farms is eventually found to have originated from commercial fertilizer, hog-lagoon seepage, or land application of hog waste, the situation recalls another criterion that Bill Haw used for siting a big hog farm. He was attracted to the highly permeable soil of northern Holt County, "a very sandy loam that can be worked one day after a one-inch rain, and on which effluent can be applied 300 days in a year."[25] In other words, when sprayed on cropground the effluent from National Farms' hog lagoons could be counted on to sink into Holt County soils rather than running off.

Competition for markets is another externality affected by industrial hog farms. Disappearing markets for small lots of hogs may have contributed to the loss of hog farms in Holt County. Between 1982 and 1997 the number of farms raising hogs dropped from 354 to 103.[26] Because the decline is similar nationwide, it would be going too far to say National Farms directly caused local hog farmers to go out of business. But the loss of small hog operations nevertheless affects a community.

Gary Olberding runs a supplies and service business for local dairies in Holt County and is former chairman of the county board. He says the loss of more than 250 small hog farms meant a loss of certain values held by people who own and work their own land. "Smaller operations had safer lagoons and put their manure on in amounts the ground could handle," said Olberding. "Their life and reputation were on the line. They had pride in what their ground looked like and had to face the neighbors."[27]

Over the years, National Farms became a better neighbor—asking for permits when it wanted to encroach on the county's right-of-way, for example, and controlling odor. Among the smaller hog producers that survive in Holt County are several who once worked at National Farms and have adapted some of the company's methods to a smaller scale that is more accommodating to neighbors.

For example, Kelly and Marge Huston feed forty-eight hundred finisher hogs in four barns on Highway 20 east of Atkinson. Both of the Hustons once worked for National Farms. After eleven years with the company, Kelly was managing a twenty-four-hundred-sow farrowing unit when he quit to start his own hog business. He says the pay and benefits at National Farms were something he couldn't have found working on a ranch. "A lot of people who had worked on ranches had no time off," says Huston. "At National, every other weekend you had off plus two weeks' paid vacation and health insurance."[28] Custom-feeding hogs for another producer gives Huston some financial stability. "We get paid a set rate," says Huston. "No matter if the market goes up or down, our cash flow is the same."

To minimize odor, rather than flushing hog waste through center pivots, Huston stores it in six-foot-deep pits beneath the barns. Twice a year, Huston cleans out the pits and transports the waste to his family's cropground, where it is disked into the soil. Lagoons lose about two-thirds of their nitrogen content to evaporation, so Huston's method retains more nitrogen for the crops. It also cuts odor to a minimum—both at the hog farm and when the manure is applied to crops.

When asked why everyone doesn't handle their hog waste in this manner, Huston says, "Lagoons are easier and they take less maintenance. If you have a lot of barns, it's cheaper to use lagoons."[29]

The Hustons and their five young children live about a half mile from their hogs. Kelly Huston's father lives across the road. If odor becomes a problem, the family will experience it first.

FURNAS COUNTY

The 4-H livestock exhibits at the 2001 Furnas County Fair include cages with elaborately feathered chickens and exotic rabbits; in long open-sided sheds are pens of sheep, beef, and swine on clean straw bedding. In the last twenty years the kinds of livestock displayed here haven't changed much, but the number has dropped with the decline in the number of farms in Furnas County—about one hundred fewer than twenty years ago.

This year about fifteen boys and girls from age nine to fourteen are showing pigs in the 4-H swine show. In the past, many 4-H youth showing hogs at any county fair in Nebraska could have chosen them from their own families' herds. But only twenty-four Furnas County farmers raise pigs, compared to seventy-two just ten years ago.[30] In talking with the swine exhibitors' parents—all farmers—I find no one who any longer raises pigs. For their children's 4-H projects, they buy show pigs from the few producers who still raise purebred hogs for show. One farmer says he expects that, by feeding a couple of Duroc pigs to market weight for the fair, his two children will understand what it takes to produce the pork chops they buy at the supermarket. I begin to think of the 4-H swine show at the Furnas County Fair as a kind of living history event, where people come to reenact and recall the past.

For many years farmer Leon Riepe has provided ribbons and trophies for the swine show at the fair. Although Riepe too has left the hog business, he continues to donate the prizes. "There aren't near as many kids in the county as there used to be," says Riepe, "and the fair isn't what it used to be."[31]

For all of his sixty-eight years, Leon Riepe has lived on and farmed Furnas County land that his great-grandfather homesteaded in the 1880s. But with his three children grown, educated, and working in city jobs, he expects to be the last member of the family on the land. "It's cheaper to educate them than to buy land or a farm operation," says Riepe. "It's awful hard work and a poor living. I feel bad being the end of the road, but I don't want them to come back. It gets harder each year." He says his children will inherit the land and can do with it what they want.

Riepe walks stiffly, as if all those years of labor on the farm have settled in his joints. He once had a farrow-to-finish operation with forty or fifty sows. "About what one person could handle," he says. But he sold the last of his hogs in 1995 because he couldn't keep up with the work any more and it was hard to get help. He also thought the price for hogs would get worse, and it did. He doesn't, however, blame low hog prices or the disappearing market on the area's biggest pork producer—Furnas County Farms, a partnership connected to Sand Livestock. Riepe says the market for small lots of hogs would likely have disappeared even if Sand wasn't in the county.

In the late 1980s, like most of rural Nebraska, Furnas County was still reeling from the latest farm crisis. Then Chuck Sand, the president of Sand Livestock Systems, Inc., of Columbus, proposed building a two-thousand-sow farrow-to-finish complex that would employ seventeen people with an annual payroll of $450,000. Annually it would produce forty thousand hogs and buy about four hundred thousand bushels of corn.[32]

Rather than concentrating tens of thousands of hogs and their manure on one site, Sand would build the operation on three sites several miles apart—one each for farrowing, nursery, and finisher. The distance between the farms would help control disease. Furnas County had a population density of fewer than eight people per square mile; Sand promised to site each farm at least two miles from the nearest neighbor to limit the impact of odor.

Sand Livestock began exploring possibilities in Furnas County at about the same time the company was fighting lawsuits in Michigan over the way its hog farms there were managed. Neighbors with complaints about odor, manure spills, and inappropriate disposal of hog carcasses had sued. Eventually the state of Michigan joined the lawsuit. In the 1989 settlement of the case, Sand Livestock admitted no wrongdoing but was barred from doing business in fifty Michigan townships and assessed twenty thousand dollars to compensate the state of Michigan for surveillance and monitoring costs at a hog confinement in Parma, Michigan.[33]

Whether or not Furnas County residents knew about events in Michigan, there was little opposition to Sand's proposal in Furnas County. Some wondered if Sand Livestock Systems, Inc., would be violating Initiative 300 by owning farmland and pigs in Furnas County. But Chuck Sand, a critic of

1-300, said his business there—Furnas County Farms—would be operated as a general partnership, which is legal under 1-300.[34]

Leon Riepe and five other residents bought an ad in local newspapers to express their alarm. They feared the loss of markets and thought the concentration of manure would pollute the water. Some of those fears have been realized.

Markets for small lots of hogs have, in fact, dwindled. For many years, farmers could sell their pigs to packing-plant buyers at the Lexington Livestock Market two days a week. Each week, dozens of farmers would sell as many as three thousand pigs in small lots. By 2002 farmers were bringing in only 150 to 200 hogs to sell one day a week, and management considered abandoning hog sales altogether.[35] Buyers for big meatpackers like IBP and Farmland get most of their hogs for slaughter from big producers like Furnas County Farms, which can guarantee thousands of hogs each week—one reason for the disappearance of hogs from local sale barns. In addition, hog-buying stations that were once scattered across the countryside, buying small lots of pigs for slaughter at the big packing plants, have all but disappeared.

There have been no environmental disasters in Furnas County with the coming of hog confinements, but there have been some spills. In the early years of Furnas County Farms' presence in the area, there were at least four manure spills from waste lagoons at the company's sites in Furnas County and in neighboring Gosper and Frontier Counties. In November 1992, for example, soil eroded around a pipe through a lagoon berm in Frontier County, allowing 85 to 90 percent of the lagoon contents to escape. The water and waste ran down a draw and filled another farmer's stock pond more than a mile away. NDEQ records show that "the excess water then overflowed threw [sic] an 8" overflow tube and proceeded south down the draw. . . . It then crossed under a county road by way of a culvert" and was stopped by a dam in another farm pond.[36] Sand Livestock, which was managing the operations for Furnas County Farms, cleaned up the mess from this spill and three other less extensive spills to the satisfaction of the neighbors and the state.[37]

Odor has been a problem for some neighbors. Several Furnas County residents who are occasionally bothered by odor told me they don't complain because family members or friends work at the hog farms. And even though Leon Riepe still objects to Sand's presence, he admits, "They feed a lot of corn and pay eight to ten cents more a bushel, no matter what price corn is going for."[38]

When big hog farms generated statewide controversy starting in 1997, Furnas County—like most other counties—eventually yielded to pressure from the legislature and adopted zoning. But unlike other counties where proposed zoning ordinances have caused heated debate, county officials said no one showed up for the hearing on Furnas County's zoning regulations—an indication that residents have come to terms with the hog operations.

Prominent Furnas County residents say Furnas County Farms is a good citizen. In testimony to the legislature's Natural Resources Committee, Alan Thomas, a banker and chairman of the Arapahoe Chamber of Commerce, listed benefits from the operations in Furnas County: $160,000 in property taxes in 1996, about $1.7 million in wages for seventy-five to eighty jobs, and a market for 25 percent of the corn grown in the area.[39]

Thomas and Beaver City mayor Steve Forbes went to Chase County to praise the company when the Chase County board considered a permit there. Forbes said Furnas County Farms was a good citizen, donating money to community causes, including new uniforms for a local baseball team. "Can I guarantee you won't smell it? No," said Forbes. "But Furnas County Farms is not the downfall of the family farm."[40] Forbes also said in a newspaper interview that with all the benefits that had come from Sand's presence in Furnas County the only question is whether the hog operation pollutes the groundwater. He said, "I just can't imagine it will, as careful as they are."[41]

Whether Sand's operations or many others in the state have polluted the groundwater is largely unknown. None of Sand's lagoons in Furnas County have groundwater monitoring wells. Before 1997 the NDEQ had never required groundwater monitoring around hog lagoons. But records from the Lower Republican NRD show nitrate levels well below the federal standard of ten parts per million in all but one of the irrigation wells around the hog operations.[42] The nitrate levels in water near the hog farms are similar to elsewhere in Furnas County, with an average of about five parts per million.

By 1997 Furnas County Farms had eight hog operations in the county—including farrowing, nursery, and finisher sites—housing about fifty-four thousand hogs.[43] In 2001 the company said its feed mill in Arapahoe bought over 1.7 million bushels of corn a year—one-tenth of the county's production. Furnas County Farms said it employed more than one hundred people with a payroll of over $2.5 million.[44]

Representatives of Sand Livestock and its related partnerships, including Furnas County Farms, declined to be interviewed for this book. As a consequence, their points of view on the topics discussed here have come from public documents and from comments made by people familiar with the companies' operations.

In spite of the benefits it apparently provides in Furnas County, the company met resistance when it tried to expand. In March 2001 the planning commission in neighboring Gosper County rejected a request from Furnas County Farms for a permit to build a twenty-five-hundred-hog finishing barn.[45] The Gosper County highway superintendent produced figures showing that his agency's share of the company's property taxes wouldn't cover the county's cost to maintain roads to the farms.[46] There had previously been spills from two of the company's operations in the county. In 1993 waste escaped through a hole in the bottom of a nursery lagoon and ran onto ground owned by the company.[47] In April 1997 a buried pipe that carried waste from a lagoon to cropground burst, and some of the waste reached a tributary of Turkey Creek.[48] The NDEQ required the company to repair the lagoon and pipe to prevent future discharges.

Word of the mixed effects of the company's hog operations in south-central Nebraska spread to other counties. About 150 miles to the west, at hearings before the Chase County Planning Commission and Chase County Board of Commissioners in 2001, dozens of people opposed Furnas County Farms' plans for a farrowing operation in northern Chase County. In Hayes County, residents objected to the company's plans to put a forty-four-thousand-head finisher on a precarious site near a stream feeding into Frenchman Creek. There was also opposition in Red Willow County to the company's plans to build there. In each case, residents feared that the hog farms would pollute the area's water and air.

Frustrated by delays and outright rejection, Tim Cumberland, Furnas County Farms partner and executive vice president of Sand Livestock Systems, wrote to state senators warning that the hog industry would leave Nebraska for more welcoming jurisdictions, taking all the economic benefits with them.[49]

Clearly, many Nebraskans' responses to proposals for hog confinements haven't rested solely upon the potential for positive economic impact—goods and services sold, jobs created, and taxes paid. Critics have pressured government officials to factor in other quality-of-life issues as well when considering whether to welcome such operations to their communities.

RESISTING THE WAL-MARTING OF AGRICULTURE

On rugged pastureland in Greeley County in east-central Nebraska stand a couple of dozen long, low, white buildings, each with a silver silolike structure at one end. This is farm country, but these buildings don't look like barns. A rank odor often taints the air. The gated entrance to the operation is guarded by a large, enigmatic sign. It says:

WOLBACH FOODS

To be a least cost producer, providing a value added product, while being environmentally and neighborhood friendly as we enhance our team members quality of life.

NO UNAUTHORIZED PERSONNEL BEYOND THIS POINT

Beyond the gate is one of Nebraska's biggest hog operations, confining about fifty thousand feeder pigs. The sign's cryptic message neatly captures elements of the conflict over dozens of such operations that sprung up in Nebraska in the late 1990s: operators promised prosperity and negligible environmental impact while shielding themselves from their neighbors' concerns.

Among Wolbach Foods' many concerned neighbors were Ron and Carol Schooley, who operated an organic farm a mile south of the hog operation. The Schooleys' lush bean fields, fragrant vegetable garden, and open pens for chickens and cattle offered a dramatic contrast to the factorylike operation north of them. The couple had always been advocates of sustainable agriculture, but until bulldozers broke ground for Wolbach Foods, the Schooleys didn't consider themselves political or social activists. Then in 1997, when they discovered the nature of their new neighbor, they led the way in trying to inform other neighbors, county boards, the legislature, and the state about what they perceived as threats to production agriculture and the environment. On a summer evening in 1998 Ron Schooley stood in his front yard and contemplated the roofs and feed bins of Wolbach Foods that now dominated his view to the north.

"It's a manifestation of where production agriculture's going today and the demise of the family farm," he said. "And I have to see it every morning from my front porch. They call themselves production agriculture, but they're a manufacturing outfit. We hear from all over that it's the Wal-Marting of agriculture, and it's inevitable, that we don't have any choices. But we do have choices."[50]

Ron Schooley chose to fight the trend, despite his own failing health. In 1998 he was diagnosed with a rare form of cancer that would take his life in May of 2000. But as a founder of Mid-Nebraska PRIDE, he spent much of the time he had left trying to persuade Nebraskans to resist the changes he saw taking place not only in his own community, but statewide.

For others, the industrialization of hog production was a logical development, in line with the increased mechanization of farming that began seriously with the invention of the tractor. Jim Pillen, who owned Wolbach Foods, often reminded critics that he grew up on a family farm in northeast Nebraska. He said those who didn't raise hogs in confinement were out of touch with reality, didn't care about making money, and shouldn't criticize those who did. Yet he acknowledged a certain loss.

"There's not a one of us who grew up on a farm who wouldn't love to be able to turn the clock back and have that way of life for our family because there's great discipline that I and my brothers grew up with," said Pillen. "You hope you can instill those values into the next generation, even though we don't have that exact way of life."[51]

Opponents of factory hog farms said the owners' focus on profit had little to do with the rural values that Pillen recalled. In comparing his own operation with Pillen's, Ron Schooley said, "They're a manufacturing outfit, I'm a farmer"—a distinction that implies powerful traditions and meaning. Grassroots activists like Schooley feared the impacts that big confinement operations would have on the environmental, social, and economic fabric of rural neighborhoods—a fabric that was held together largely by the traditions of family farming.

A characteristic of a family farm is that typically the owner lives on the farm, has invested capital in the operation, and is responsible for day-to-day labor and management. More often than not, members of the family participate in the daily work and share in the rewards.

All of Nebraska's biggest hog confinement operators live at some distance from their livestock. Jim Pillen's offices at Progressive Swine Technologies in Columbus—sixty miles from Wolbach Foods—are decorated with aerial photographs of his widespread operations. Pillen hires managers to look after the day-to-day work and spends most of his time on executive tasks like dealing with lenders and buyers and testifying at zoning hearings to defend his business.

In a 1998 interview the mayor of Orleans, Tom Thomas, observed the differences between livestock operations that are managed by hired help and farms run by families who live on them: "Most of these farmers around here were born and raised in the area. Their grandparents settled this country. They expect their children to be on that land. They have ties to community. They're in our churches, they support our chamber of commerce, they're the ones that work with our Little League baseball and they work with Scouts. . . . Certainly they have an economic interest because they want to make a living for themselves and their family, but they have ties to this

Ranchers Jim Lawler, Ron Lage,
and Barb Rinehart (*left to right*) fear
that hog waste will pollute Sandhills
water. (Credit: Carolyn Johnsen)

Max Waldo (*left*) and his father, Willard, operate Waldo Farms, which has raised Duroc hogs without controversy for more than one hundred years. (Credit: Carolyn Johnsen)

top, right, Mabel Bernard said she couldn't plant enough flowers to overcome the odor of thirty-six thousand hogs near her family homestead in Chase County. (Credit: Tina Kitt)

Wayne Kaup has a strategy for increasing the value of his land with hog waste. (Credit: Carolyn Johnsen)

Top, left. Brian Mogenson by the lagoon at one of his hog farms in Antelope County. (Credit: Carolyn Johnsen)

Bottom, left. Aaron Spenner and friend on his family's hog farm near West Point. (Credit: Carolyn Johnsen)

Bob Spenner's hog pasture near West Point is part of an operation that raises swine for special markets. (Credit: Carolyn Johnsen)

Left, top. As in other confinement operations, finisher pigs at the University of Nebraska hog farm near Mead are raised indoors on concrete floors. (Credit: IANR News and Publications)

Left, bottom. One National Farms site in Holt County. (Credit: Richard Whiteing)

Ron Schooley joins other factory-farm opponents at a State Capitol rally. (Credit: Bill Batson, *Omaha World-Herald*)

Top. A sow and her litter at a Progressive Swine Technologies farrowing operation. (Credit: Barrett Stinson, *Grand Island Independent*)

Right, top. Hog farms quickly became a focus of Nebraska news starting in 1997. (Credit: Paul Fell)

"If you can't beat 'em, move." Drawing by seven-year-old Nicholas Mullanix, whose family lives near forty-eight thousand hogs.

Elaine Thoendel (*left*) and Donna
Ziems are persistent critics of the
Nebraska Department of Environ-
mental Quality. (Credit: Bill Batson,
Omaha World-Herald)

community that give them a greater sense of responsibility. When we have outside interests that come in whose only goal is profit, as soon as that operation does not return a profit, the assets will be put back on the market."[52]

Objection to absentee ownership of farmland has long been a theme in the writings of Marty Strange, a founder of the Center for Rural Affairs in Walthill. He writes, "The natural aversion agrarian communities everywhere seem to have for absentee ownership of land is . . . more than the parochial jealousies of a backward, suspicious people. It is based on their heritage of experience that when land accumulates in the hands of speculators, people who depend on working the land for their living are sure to suffer. . . . Some people value community and neighbors too much to willingly sacrifice them to the market. For them, farming is not merely an occupation or an opportunity, but an identity. Their farming values are not merely economic, but social and cultural as well. For these socially minded farmers, farming in America is like swimming against the tide, but they do it their way because their values require it of them."[53]

One of those rural values is trust. In Nebraska the customary high level of trust among neighbors eroded with the surge in pork production that began in 1997. Rural people have always been alert to their neighbors' comings and goings, but the neighborhood watch intensified around big hog farms. County clerks observed a flurry of public interest in deed transfers as land was acquired for hog operations.

The expansion in hog production that occurred in Nebraska in the late 1990s was accompanied by promises of prosperity for communities that craved economic growth. For decades rural citizens and policy makers alike have mourned the loss of population in Nebraska's rural counties and have sought ways to reverse the trend.

Rather than being a solution to the problem, factory hog farms apparently contributed to it. There is a correlation between population loss and increased hog production; that is, in rural counties where hog production increases, population declines.[54] One reason is that confinement feeding requires less human labor. It replaces workers with mechanization, resulting in the loss of two or three independent farmers for every confinement job created.[55] A common employment standard in the industry is one full-time person handling four or five twelve-hundred-head finishing buildings.

Another likely cause of population loss is the deteriorating quality of neighbors' lives that occurs when hog operations are poorly managed. Farm families are accustomed to occasional odor from livestock, but many have found unbearable the stench generated by thousands of hogs confined in a limited space.

Rural folk who expressed their fears of these kinds of effects on rural life and culture were often dismissed as being emotional rather than sensible and pragmatic. Annette Dubas, a founder of Mid-Nebraska PRIDE, had an answer to that charge.

"I used to defend myself about being emotional," she said. "But I don't any more, because this is an emotional issue. And when you feel like your lifestyle is being threatened, you have a right to be emotional."[56]

The factory hog farms that established themselves in rural Nebraska at the turn of the millennium did change some cherished elements of agrarian life in unwelcome ways, but a cooperative grassroots effort to shape public policy left positive marks on other aspects of rural culture—civic and social life. Activists formed new friendships, ameliorating feelings of helplessness with a shared sense of accomplishment and a new appreciation of their role in government.

Chuck Hassebrook, director of the Center for Rural Affairs in Walthill, has long been an advocate for traditional family farms. Of the grassroots groups that formed around hog farm issues, Hassebrook says, "I think they just make democracy work better because they force government and policy makers to respond to critical concerns of average people—of ordinary citizens that oftentimes the most powerful groups and players don't respond to."[57]

The history of modern environmentalism is largely a story of ordinary people pushing for change while governments reluctantly follow behind.

– Mark Hertsgaard, *Earth Odyssey*

The Marshal Comes to Dodge

In Elaine Thoendel's kitchen, stacks of paper and bulging manila file folders compete for limited space with pots and pans, detergent bottles, and toaster. Although Elaine and Dennis Thoendel have seven sons, ages seven to seventeen—all living at home—the most heavily used appliance in this kitchen may not be the refrigerator or stove but the copy machine that occupies one end of the dining table. The copier is a necessary tool in homeschooling the children. It's also essential to Elaine Thoendel's scrutiny of big hog operations in her neighborhood and of the state agency charged with regulating them.

Since 1997 Thoendel, a small dairy farmer, and her friend Donna Ziems, who raises a few hundred hogs each year, have added to their farm and family obligations the self-imposed task of monitoring how Premium Farms and the Nebraska Department of Environmental Quality operate. In 1997, when Brian Mogenson of Premium Farms began buying land in Holt and Antelope Counties, the women decided, "If they're coming, we've got to make sure we know what the laws are so they follow them."[1] Since then they've become citizen experts on Title 130, the fifty or so pages of regulations that govern livestock operations in Nebraska.

Given any encouragement, the women expound at length on the daily production of volatile solids from sows, nursery, and finisher pigs; the amount of water needed to flush underfloor pits; the seepage rate of live-stock waste lagoons through various soil profiles; the agronomic rates for nitrogen and phosphorous applied to corn, alfalfa, and soybeans; and the difference between slurry and sludge.

The two women had previously been on opposite sides of a school-closing issue. But in this effort they became a team heading up an organization they call "Nebraska Worth Fighting For." To anyone who would listen, the women explained their concerns about the NDEQ's enforcement practices. Ziems often took her own copy machine to the NDEQ office in Lincoln and spent hours there in a small glass-walled room duplicating the agency's records on Mogenson's hog farms in Holt and Antelope Counties.

Mogenson said he felt "hounded" by the two women. To him, the rows of long, white buildings that housed his hogs, the glassy surfaces of the big lagoons, and the neatly mowed grounds were not only beautiful but were exactly what the economy in northeast Nebraska needed.[2] Doug Rowse, a manager at the Central Farmers Co-op in Elgin, agreed. He said the co-op modernized its feed mill to handle production for Premium Farms' nine finisher units, which required 2.25 million bushels of corn and fifteen thousand tons of soybeans a year. By 2001 the feed mill was running twenty hours a day; Rowse hired seven additional people to handle the increased trucking, marketing, and paperwork.[3]

But Elaine Thoendel and Donna Ziems were not persuaded. The women scrutinized Mogenson's operations—not only on paper but in person. Standing on the public right-of-way bordering Premium Farms construction sites, they took date-stamped photos of heavy equipment moving earth and pouring concrete before the state had granted permits for construction. They studied geological survey maps and soil surveys of the Elkhorn River watershed and learned how to estimate when seepage from hog waste lagoons would reach groundwater and the streams that feed the river. They worked tirelessly to explain what they knew to state and county officials, members of the legislature, and the press. It was a four-hour drive to the capital in Lincoln; they drove the distance dozens of times to keep the pressure on.

Ziems even did surveillance of a kind on Premium Farms. Neighbors had seen trucks with Oklahoma license plates loading and unloading hogs at Premium Farms sites. Ziems suspected some kind of arrangement between Brian Mogenson and Seaboard Farms, Inc., which operates a big packing plant in Guymon, Oklahoma. In states that permit nonfamily corporations to be engaged in agriculture, it would have been legal for Seaboard to own the pigs in Brian Mogenson's hog sheds. But Initiatiave 300—Nebraska's anticorporate farming law—made it illegal.

One winter evening in 1999 Ziems heard that Oklahoma trucks were loading pigs from a Premium Farms facility. She followed one of them south toward Kansas. At one stop, Ziems asked the driver for veterinary papers on the pigs. She got a good enough look at the papers to copy information that pointed to ownership by Seaboard—evidence that was later used by the Nebraska attorney general to accuse Seaboard Farms of violating I-300.[4]

Donna Ziems and Elaine Thoendel said they did these things for their children, and to protect the air, water, and land from exploitation by those who would use them for short-term profits. "There's no future if we don't save this," said Ziems.[5] But livestock producers who were targets of this kind of activism said it put at risk their investments in property and livestock and their very right to make a living off the land.

Opposition to hog confinements in Nebraska—as in other states—typically began with a response to a local development. Elaine Thoendel and Donna Ziems wanted to protect nearby Clearwater Creek, the Elkhorn River, and their groundwater from contamination. The leaders of Mid-Nebraska PRIDE were concerned for their groundwater, the Cedar and Loup Rivers, air quality, and the future of their family farms. In north-central Nebraska, concern for a trout stream catalyzed opposition.

In Rock and Brown Counties, Long Pine Creek cuts a deep canyon through pasture, hayfields, and irrigated corn and soybeans. For travelers along U.S. Highway 20, the sudden dip into the piney woods on the banks of this pristine stream comes as a surprising and welcome break in the five-hundred-mile monotony of cropground that dominates the highway scenery from Dubuque on Iowa's eastern border to Valentine, Nebraska. The steep, wooded canyons, the clarity of Long Pine Creek, and the abundant trout that inhabit it have attracted tourists for decades.

Campgrounds, rental cabins, and hundreds of vacation homes have been built along the banks of Long Pine Creek. They're owned by retirees and business people from as far away as Grand Island, Omaha, and Lincoln and by local residents whose families have spent leisure time in this pristine natural area for generations. The clear, cold creek rises in the Sandhills in southern Brown County, flows along the border of Rock County, and joins a tributary of the Niobrara River about thirty-five miles to north. It's fed by spring waters—in particular, seven springs that provide drinking water for the village of Long Pine.

The water is so pure that in 1997 a small, locally owned company started bottling it and selling it as Seven Springs water throughout Nebraska and into South Dakota and Kansas. With ten employees, the Seven Springs bottling plant was the only manufacturing firm in all of Rock County— one indication of the scarcity of jobs outside of agriculture.

Rock County citizens had tried to attract other businesses to the area. Then in 1998 Brian Mogenson began acquiring options to buy land in Rock County. He announced plans to build at least six hog farms, each housing fourteen thousand swine—all in the watershed of Long Pine Creek. A local real estate agent hailed the economic benefits of the project. A Rock County commissioner said the county's new zoning regulations and state laws would protect the area from pollution. Mogenson recalled his supporters telling him, "We know that General Motors is not gonna build a factory here in Rock County. We don't have no iron ore under the ground. We've got agriculture."[6]

But a small group of farm and ranch women—most of them lifelong residents of the area—said Mogenson's planned operations were more like factories than agriculture, and they didn't want millions of gallons of hog manure anywhere in their pristine watershed. They were offended by a real estate agent who urged them to sell land to Premium Farms, offering them several times the going rate per acre.

Loranda Daniels-Buoy, whose parents had a hundred-sow farrowing operation on their farm when she was growing up, said, "Having experienced hogs on a small scale I know what big-scale hogs could be—what it would smell like and what it could do to our water." Loranda's sister-in-law, Lynda Buoy, who ran up a three-hundred-dollar phone bill in researching the issues, objected to the way confinement operations treat animals.[7] These rural women found allies among city people who owned vacation homes on Long Pine Creek. The State Game and Parks Commission received dozens of letters opposing Premium Farms' plans. The commission director said they weren't form letters, but "They are people writing from their heart expressing strong concern."[8]

Calling themselves CARE—Citizens for Air, Resources, and Environment —the group attracted about fifty members and collected about nine thousand dollars to publicize community meetings, buy ads in the local paper, and print flyers citing scientific studies, news stories, and other information they'd found in their own effort to become educated on the issues. Their survey of Rock County voters showed that 73 percent wanted to keep the hogs out.

Brian Mogenson was baffled by the intensity of opposition to his plans. "We're building the finest facilities that we can build," he said. "Everybody in the hog industry's impressed with my buildings. It's hard enough and risky enough trying to raise thirty-dollar hogs without all this turmoil going on."[9] Another factor contributing to the hostility toward Mogenson was suspicion about who backed his ventures in Nebraska. Premium Farms' construction was financed by a thirty-million-dollar loan from a Delaware limited-liability company that wouldn't reveal its investors.[10] There were rumors that the money came from A. J. DeCoster, the nation's twenty-third-largest pork producer.[11] DeCoster raised both pigs and chickens and had numerous environmental and labor violations stretching from Maine to Iowa. Mogenson and his father, Harry, had erected swine buildings and feed mills for some of DeCoster's Iowa operations, but Brian Mogenson said DeCoster wasn't involved in the Premium Farms enterprises in Nebraska. Although Mogenson's critics, their attorneys, the NDEQ, and journalists investigated, no one conclusively tied DeCoster to Premium Farms. Rumors persisted.

In May 1999, when the Rock County board considered Mogenson's applications for permits, none of his supporters showed up. Mogenson himself didn't make it to the meeting. Mogenson's attorney and engineer touted the environmental safety of their lagoons and the economic benefits that eighty-four thousand pigs would bring to Rock County. But farmer and CARE member Geri Kuchera named a dozen families whom the hog farms would affect. "We have invested our money and our hearts in our farms and ranches," said Kuchera. "We are proud to work hard to leave something for our children to inherit." She said that if the hog farms were allowed in the neighborhood, "they will think first of their profits and not of the community and they will leave a trail of waste for all of us to inherit and clean up."[12]

In a unanimous vote, the Rock County board turned down Mogenson's applications and the crowd applauded and cheered. One newly elected commissioner said of his constituents, "They put us under a lot of pressure and kept us under it."[13] But Premium Farms' attorney said, "I think that, as with most things, those who oppose are more vocal and more organized, and I guess I was not surprised at the makeup of the audience."[14]

With the door shut to his Rock County plans, Brian Mogenson concentrated on Antelope and Holt Counties, where Donna Ziems and Elaine Thoendel would add to their list of complaints an allegation that Mogenson was violating the state's ban on corporate farming.

As big pork producers eyed other unzoned areas of the state, protestors organized there as well. When Enterprise Partners—a company connected to Sand Livestock—applied to the NDEQ to build hog farms in Perkins County, neighbors organized Save Our Rural Resources (SORR) to monitor the operation and lobby the legislature. SORR member Jim Hanson, whose family had farmed in the area since 1889, expressed the fears felt by other hog-farm critics. "I might be able to put up with the struggle of farming," said Hanson. "But I can't put up with poisoned water or the air stinking so bad that you wouldn't want to stay."[15]

In Hayes County, when Sand Livestock and Furnas County Farms broke ground for a forty-four-thousand-head finishing operation above Stinking Water Creek, yet another organization formed to resist. They called themselves Area Citizens for Resources and Environmental Concerns—ACRES.

During the public comment period on the NDEQ permit for the site, ACRES sent to the NDEQ a list of what they believed were Sand's and Furnas County Farms' previous failures to comply with environmental law in Nebraska and Michigan. The details—most of them drawn from public records—included operating without permits, manure spills, and illegal disposal of swine carcasses.[16]

ACRES asked the NDEQ to deny a permit for the new operation. In its response to the complaints, the NDEQ acknowledged that many of the problems had occurred but that Furnas County Farms and Sand had cooperated by cleaning up spills and eventually complying with regulations. The agency said, "All violations or alleged violations were responded to by the applicant in a timely manner or were of a sort that are noted and corrected by the agency through its inspection process."[17]

ACRES also pointed out Sand's problems in Michigan; the NDEQ said the company had admitted no wrongdoing in Michigan and that the judgment issued there was more than five years old, so the bad-actor clause of Nebraska's livestock waste law didn't apply. On 7 February the NDEQ granted Furnas County Farms its Hayes County permit.

Two weeks later, Sand Livestock, Furnas County Farms, Chuck Sand, and Tim Cumberland sued ACRES' leaders for defamation. The lawsuit accused them of publishing information that was motivated by malice and designed to injure the company.[18]

ACRES co-president Char Hamilton said, "We're just a bunch of people concerned about the environment, that's all we are."[19] The expense of defending themselves in a lawsuit added stress to the group's intense concerns for the environment. Tina Kitt, editor of the local newspaper, the *Wauneta Breeze*, said the lawsuit wasn't popular in the community. "People are offended by it," said Kitt. "It's just over the top—even if they support business."[20]

ACRES' attorney called the litigation a "SLAPP" suit—for "strategic lawsuit against public participation"—illegal in Nebraska and most other states. In October 2001 ACRES received some big-time help from the Waterkeeper Alliance of White Plains, New York, a national coalition of law firms that had filed lawsuits against big hog farms in other states alleging violations of environmental law. Robert Kennedy Jr., who headed the group, said Sand's lawsuit against ACRES was a "bullying tactic intended to silence constitutionally protected speech and divert attention from Sand's extended history of environmental lawbreaking. These bullying tactics have a chilling effect on the democratic process."[21] Capturing the flavor of a Western showdown, Kennedy added, "Now Sand knows that the marshal has come to Dodge." Tim Cumberland, Sand's executive vice president, said, "As far as bullies go, there's a lot more bullying coming from the East Coast than from the Midwest."[22]

Other organizations with experience in environmental issues have also intervened on behalf of Nebraska activists. As part of its national campaign against hog confinements, the Sierra Club assigned a worker to Nebraska to help grassroots efforts to organize and lobby. The Center for Rural Affairs provided an experienced organizer, Nancy Thompson. Of the grassroots activists, Thompson said, "They had no money, no organization, and no sense of what needed to be done. But they learned quickly."[23] Thompson helped them focus on how to begin the zoning process and organized farm-lobby days at the legislature. They learned how to write press releases, hold news conferences, and testify at public hearings.

Leaving themselves open to charges of being influenced by "out-of-state agitators," Nebraska's opponents of industrial hog farms also sought advice from experienced activists in other states.[24] It was easy to do, given the multitude of websites devoted to the hog-farm controversy. Sue Jarrett, a fourth-generation rancher in Yuma County, Colorado—just over the Nebraska state line—acted as a kind of cheerleader for Nebraska activists, urging them to stick together and educate themselves both on the issues and on how their government works. Jarrett drew heavily upon her own experience living near a 400,000-hog Seaboard operation.

Karen Hudson, an Illinois farmer, started Families Against Rural Messes (FARM) in the mid-1990s when a 1,250-cow dairy confinement was built near her farm. Like many such activists, Hudson expanded her efforts beyond her home state. She advised other groups to adopt a catchy name and to make signs to capture the media's attention. Some of Hudson's favorites are "Pigs Don't Vote," "Speak now or forever hold your nose," and "Illinois: Land of Stinkin'" which includes a profile of Abe Lincoln with a clothespin on his nose. Having failed to stop the dairy, Hudson settled for monitoring its compliance with environmental laws. "We're kind of like detectives out here," said Hudson. "Or poop police."[25]

In fact, like environmental agencies in other states, the Nebraska Department of Environmental Quality relied heavily upon citizen "poop police" to help enforce state laws regulating livestock. Citizen watchdogs kept NDEQ's phone lines humming with complaints and tips. Acknowledging the importance of information from citizens, one NDEQ official said, "It's not possible to be everywhere at once, and nobody wants that large a government interfering in their lives."[26]

Neighbors were at odds over hogs. At county zoning hearings, this split in opinion was apparent as county board chairmen gaveled down noisy crowds who shouted in response to public hearing testimony or to the board's actions regarding a local hog farm. The "hog wars" ignited a new interest in participating in government and politicized rural Nebraskans who had never before been parties to controversy.

Before he began prodding the legislature for more stringent oversight of hog farms, SORR member Jim Hanson—in his mid-40s—observed that his only previous visit to the State Capitol had been as a teenager on a high school field trip.[27] PRIDE member Carol Schooley said, "This is kind of embarrassing to admit, but we didn't know how our government worked."[28] They learned by testifying at public hearings, writing letters to senators and the governor, scrutinizing public records, and running for office. Annette Dubas of Mid-Nebraska PRIDE became the chairwoman of the Nance County Planning and Zoning Commission. Donna Ziems was elected to the Holt County board, defeating a long-time incumbent.

Like flies biting at the ankles of legislators and bureaucrats, the hog-farm opponents wouldn't go away. Most of them were farmers, which discouraged their critics from trying to discredit them for not knowing anything about agriculture.

Farmers often disagree over policy, a fact perpetually reflected in the national discourse on the federal farm program. Even so, policy makers were surprised that so many Nebraska farmers opposed industrial hog farms and wondered why anyone would waste their energy trying to bar what, to many, seemed inevitable. Some suggested that farmers are naturally conservative and resist change. Others said hog-farm critics were merely jealous of neighbors who had figured out a way to make money from hogs. There were other reasons for the resistance.

Nebraska has a long tradition of people with close ties to the land, eloquently reflected in the writings of Willa Cather, Mari Sandoz, and Wright Morris and in the work of contemporary authors. Their fiction, essays, and poetry reflect the love of a land and a way of life hard-fought for.

Many Nebraskans who have opposed industrialized hog production live on farms or ranches that have been in their families for generations. They carry memories of a hundred years or more of the efforts spent to bring their enterprise to where it is today. For example, Mabel Bernard mingles her objections to the stench of thirty-six thousand hogs near her home with proud stories of her in-laws' struggles with the weather, buffalo, and Indians and their survival in the harsh, yet forgiving, prairie.

Nebraska's anti-hog-farm activists have had models to follow in their attempts to influence government. Hundreds of rural people carried Initiative 300 petitions in the early 1980s. Throughout the 1990s a small group of rural activists opposed plans to store low-level nuclear waste in Boyd County. They educated themselves on the hydrology of the area and on the nuances of state and federal law. At great personal expense, they hired their own experts and lawyers to represent them at meetings of the multistate group that planned to store nuclear waste in Boyd County. To dramatize his opposition, one resident of Boyd County went on a hunger strike.

In Dakota City, where an IBP beef plant slaughtered five thousand cattle each day, residents organized to persuade government officials to do something about the rotten-egg odor of hydrogen sulfide that often invaded their homes and made them ill. The plant was emitting as much as a ton of hydrogen sulfide each day.[29]

For ten years, Citizens for Environmental Stewardship, led by Linda Modlin of Dakota City, educated themselves on federal and state air-quality laws, learned what to look for in IBP's records at the NDEQ and the EPA, and pestered federal and state bureaucrats and the state legislature. Even as their property and their health were damaged by daily exposure to hydrogen sulfide, they tempered their rage and frustration. They channeled their energies into persistent lobbying and gathered evidence to make their case.

Due largely to their persistence, the U.S. Department of Justice and the EPA filed a lawsuit in 1999 to force IBP to comply with air quality laws. In October 2001 IBP settled the case by agreeing to pay $4.1 million in civil penalties related to violations of the Clean Air Act and other environmental laws at its Dakota City plant.[30]

Linda Modlin didn't live to see the outcome of that lawsuit. Her legacy is a state regulation that limits the amount of hydrogen sulfide allowed in the air. The gas is a major component of the emissions that arise from hog waste.

Most big hog producers who were the focus of grassroots discontent in Nebraska were satisfied to dismiss critics by calling them "emotional" and by accusing them of failing to consider sound science in their calls for more regulation. For one clergyman, the dispute was more personal. Father Richard Whiteing, a Catholic priest in Fullerton, often accompanied PRIDE organizers to their meetings around the state, where he spoke about the church's doctrine of economic justice. "The catechism of the church talks about the morality of one person making an excess profit at the expense of another person's livelihood," said Whiteing. "And that's what seemed to me was going on with the hog business, that the large producers were swallowing up the whole economy and controlling the economy and hurting smaller farmers."[31]

Both at public meetings and from his pulpit, Whiteing denounced what he considered the moral consequences of highly concentrated animal feeding. In a letter to the *Columbus Telegram*, Whiteing chided big hog producers for being greedy. He wrote, "Sometimes in our capitalist society we mask greed by calling it progress. Mega hog confinement operations are not progress. . . . Perhaps if those who are investing in these operations could take a look beyond their desire to make a dollar, they could see the risk they are placing all of us in."[32]

Tim Cumberland of Sand Livestock Systems replied in a letter a few days later, "Why is it that some people who depend on the charity and hard work of others for their livelihood seem to be the first ones to attack this country's system of free enterprise and capitalism?"[33] Cumberland suggested that the archbishop of the Nebraska diocese should reconsider asking for donations for "such great charitable works as fixing the ceiling of the Cathedral dome" if he was going to permit Father Whiteing to oppose legitimate agricultural enterprise. The exchange continued in a couple more letters until the *Telegram*'s editor told both men no more of their letters on the subject would be published.

Chuck Sand—a Catholic himself—wrote to Archbishop Eldon Curtiss complaining about the outspoken priest.[34] Sand asked whether Whiteing, in describing big hog farms as "immoral," represented the official position of the Catholic Church. In his reply, the archbishop defended Whiting. Curtiss said the church had no objections to big hog operations if they acted justly in local communities and used natural resources responsibly.[35] Shortly afterward, Father Whiteing was reassigned to a parish in Omaha and then to Jackson, Nebraska. He continued to write and speak about moral economic choices.

Although men have had an important part in the movement for tighter regulation of hog farms, women have been most prominent—in Nebraska as in other states. You could call it the "Erin Brockovich" syndrome, after the unemployed single mother in California who helped bring Pacific Gas and Electric to its knees for polluting a small community's water with toxic levels of hexavalent chromium.

Sociologist Kendall Thu has spent much of his career studying the impacts of the concentrated feeding of pigs on rural communities. He says the activism of rural women in this issue fits their traditional roles. "We know in many surveys of farm families, women are the caretakers of the kids and spouse, and the health care providers for family and household," says Thu. Also, rural women aren't as likely as their urban sisters to have jobs outside the home and so have an opportunity to be leaders. Thu says, "I also think women in rural areas tend to have a richer social network and somehow can use that to capitalize and mobilize like-minded folks in opposition—that is, mobilize the social resources."[36]

The influence of these groups was often felt by members of the legislature's Natural Resources Committee. Lincoln senator Chris Beutler—committee chairman when the controversy began—said citizen groups were typically less focused and less strategic in their efforts than professional lobbyists. Beutler said, "If you talk to the citizen lobby, they're going to be irrelevant about three-quarters of the time. And telling their individual stories about how they got involved with the issue and what's happening in their own little piece." But Beutler added, "Obviously there wouldn't have been anything done about this situation if folks in the rural areas had not made an issue of it."[37] Beutler gave rural activists credit for the sudden enthusiasm for county zoning, which the legislature had unsuccessfully advocated for years.

In 1999 Beutler lost the chairmanship of the Natural Resources Committee to Ed Schrock, a farmer from Elm Creek. After a couple of years at the center of the debate over public policy related to livestock, Schrock observed of the activists, "It's a hardcore group out there, dedicated to the cause. . . . I think it's genuine concern over the resources of the state, a combination of concern about the environment, and concern about the way farming's going." He said some had "an alarmist mentality, but they've got a right."[38]

Some critics called the activists "CAVE" people—Citizens Against Virtually Everything. But a 1998 poll showed that the several hundred rural

Nebraskans who organized to resist mega–hog farms represented opinions held by a majority of their neighbors.

Each year the Nebraska Rural Poll asks people in the state's eighty-seven rural counties for their opinions. In the 1998 survey of 4,196 rural Nebraskans, the poll found that 54 percent believed that—even if properly managed—large-scale pork operations damage the environment. Fifty-seven percent believed that, compared with large-scale pork operations, smaller pork farms were better for the state and local economy. Seventy-nine percent thought large-scale operations owned by local farmers were better for their community than those owned by outside investors. Rural Nebraskans' concerns about big hog operations increased if the units were closer to their homes.[39]

It is important to note that groups like ACRES, SORR, PRIDE, and Nebraska Worth Fighting For may have slowed down the spread of industrial hog farms in Nebraska, but they seldom stopped them altogether. Nebraska is, after all, a state friendly to agriculture. Instead of banning mega–hog farms, counties typically adopted zoning ordinances that accommodated local ideas about the kind of farming that best fits the locality and the values of the people who live there.

Confusion arises when we do not acknowledge that environmental disputes involve both private goods that are allocated by the market, and public goods, such as open space or clean air, that are allocated by politics.

– Thomas Dietz, "Thinking about Environmental Conflicts," George Mason University, 2001

Pork Tenderloin at the Capitol

One day in March 1998 the savory aromas of sage and pork filled the hallways of the Nebraska State Capitol. In a room on the west corridor, tables were laid for a feast of pork tenderloin with dressing and a glaze of mangoes and peaches, new potatoes, asparagus spears, and Black Forest cake. Hosting this elegant meal were about sixty members of the Nebraska Pork Producers Association; their guests were about a dozen state senators. The NPPA senators' luncheon was an annual event, but this year more was at stake than the usual goal of reminding policy makers of one of the state's major economic forces. With one day of debate on the Livestock Waste Management Act under their belts, the senators would make a timely acquaintance with some who would be most affected by the bill.

At the event was eighty-six-year-old Willard Waldo of DeWitt, a founder of the NPPA. The Waldo family, who had farmed without controversy in Saline County for more than a century, owned the largest herd of purebred Duroc hogs in the nation. Willard Waldo had raised hogs all his life, from the early times on dirt lots and pasture to more recent total confinement. He had also served a brief time in the legislature. Waldo understood that many senators felt conflicted by the sudden influx of big hog farms into Nebraska; they were torn between two sets of constituents—those who saw the trend as an economic opportunity and those who feared environmental disaster.

"There are a few of these great big operations—not necessarily in Nebraska—that have been ruthless," said Waldo. "They've been after the money. They haven't paid much attention to waste and they haven't been much concerned with what the pollution might do to people downstream."[1]

Also at the event was Roland Nabor of Utica. A hog farmer for fifty years, Nabor ranked among the smallest in Nebraska, raising about 150 hogs a year with his son. "It's a long way from what confinement is today," said Nabor. "We use a pitchfork and a manure spreader."[2] The NPPA hoped to persuade the legislature to limit the impact of new laws on small farmers like Nabor who couldn't afford costly environmental engineering—and whose operations posed less of a risk to the environment than those with thousands of hogs.

Lincoln senator Chris Beutler, who attended the NPPA lunch, said it served the purpose of "signaling to people that though you may not be on their side today, you may well be on their side tomorrow, and it's not a personal thing. I'm just another guy looking for what I think is the best public policy."[3]

Interest groups and their lobbyists were more than willing to suggest policy. In 1998 the state's four biggest pork producers together spent about $112,000 on lobbying. In 1999 that figure dropped to $70,200—an indication, perhaps, that the largest producers got most of what they wanted from the 1998 legislature. The Nebraska Pork Producers Association had traditionally relied on their own staff and farmer-members for lobbying. But with unprecedented attention being paid to the pork industry in 1998 and 1999, the NPPA hired its own lobbyist for a total cost of thirteen thousand dollars.[4]

Some senators argue that lobbyists serve a valuable function by providing information about complex issues. Frequently that information is delivered out of the public eye with little opportunity for opposing points of view to be aired—except in public hearings. In 1997 and early 1998 representatives of Progressive Swine Technologies, Sand Livestock, and Bell Farms had testified along with the NPPA and other interest groups at hog-farm hearings.

But by 1999 those who spoke for the biggest pork producers—the ones who had caused the general alarm—were noticeably silent at legislative hearings. The biggest producers and their lobbyists were using a different method of influencing policy—through personal contacts with senators and the governor.

In July 1998 several pork and dairy producers, two state senators, and NDEQ officials met with Governor Nelson to resolve a conflict over the NDEQ's delays in issuing construction permits for waste facilities. At the meeting, the agency agreed to make exceptions and to allow construction to begin without a permit so that producers could get buildings up and enclosed before winter. This exception was called a "hardship variance." It would be granted with the understanding that the producer might not be permitted to operate if, once the NDEQ had time to inspect the operation, the agency decided it was improperly built.[5] The NDEQ issued at least a dozen variances in 1998 and 1999 and never denied an operating permit.

In October 1998 Governor Nelson again intervened in the tense relationship between the NDEQ and livestock producers who were trying to expand across the state. Present at a meeting with the governor were three state senators: Chris Beutler, chairman of the Natural Resources Committee; Roger Wehrbein, chairman of the Appropriations Committee; and Agriculture Committee chairman Cap Dierks. Aides for Senators Stan Schellpeper and Elaine Stuhr attended. Also present were agriculture director Larry Sitzman; Beemer pork producer Stan Ortmeier; Chuck Sand, two Sand employees, and his lobbyist, Trent Nowka; former state senator Loran Schmitt, lobbyist for Bell Farms; Jim Pillen's lobbyist, Bill Mueller; Nebraska Pork Producers president Phil Hardenburger; and representatives of the Nebraska Cattlemen, Mid-America Dairymen, the Farm Bureau, and the Nebraska Bankers Association.[6] Many of those present had publicly urged the NDEQ to speed up its permitting process to enable the confinement-building frenzy to proceed. The NDEQ's Randolph Wood and three members of his staff were also there.

A memo that Wood wrote to Governor Nelson contains a record of the meeting. The memo makes it clear that the NDEQ was under pressure to ignore the law and its own regulations so livestock producers could speed up construction.[7] Wood, who was appointed by Governor Nelson, must have felt enormous pressure at this meeting. In his memo, Wood wrote that he was dismayed at Chuck Sand's allegations that the agency didn't know what it wanted in permit applications. Wood wrote, "After I calmed down, the staff said that they disagreed with Chuck's comments, but did not want to be confrontational in public or in front of you."[8]

It isn't clear whether Nelson actually called any of the shots at the NDEQ, but merely having him present while livestock producers challenged the agency's management would have been intimidating to NDEQ staff. Mike Linder, who succeeded Wood as NDEQ director, said he didn't know what happened at that meeting because he wasn't there. "But it's usual for citizens to complain to their elected representative" when they have a quarrel with a regulatory agency, said Linder.[9]

As a longtime supporter of political candidates, Chuck Sand was more than an ordinary constituent. Twice he was the campaign manager for Virginia

Smith, who represented Nebraska's Third Congressional District from 1975 to 1991. Federal Election Commission records show that between 1990 and 2001 Chuck Sand individually donated at least $26,000 to the Republican Party and to candidates for Congress and president. With a total contribution of $24,550, Sand Livestock Systems ranked nineteenth nationally among all livestock-industry contributors to federal election campaigns in the 2000 election cycle.[10]

In the eight years that Ben Nelson was governor of Nebraska, his relationship with Chuck Sand developed during several international trade missions, on which either Sand or one of his employees was present. Nelson's February 1999 list of gifts shows a "flight to California and lodging" paid for by Sand Livestock in route to a "China summit."[11] For many years Sand has had a big hog operation in southern China.

Both Sand and Nelson are avid hunters. Among the dozens of hunting and fishing trips for which Ben Nelson accepted accommodations or travel expenses during his two terms as governor, at least six were donated by Sand Livestock.[12] At least twice—first as a candidate for U.S. Senate and then as senator—Ben Nelson has taken an African safari with "friend Chuck Sand."[13]

At the end of Nelson's two terms as Nebraska governor, his agriculture director, Larry Sitzman, took a job in international marketing with Sand Livestock. Trent Nowka—one of Governor Nelson's top advisors—went to work for the lobbying firm that represented Sand at the legislature.

Chuck Sand's contacts with elected officials continued during Governor Johanns' administration. In the 1998 Nebraska elections, Sand and his family, Sand Livestock Systems, related partnerships, other Sand businesses, and officials of those companies gave more than $42,000 to candidates for statewide office.[14]

Mike Johanns, who promised to appoint a new NDEQ director who would be more sensitive to the economic importance of the hog industry, used Sand's plane several times during his campaign for governor. Other Republican candidates for statewide office also used the plane. Campaign finance records show that Don Stenberg (running for attorney general), Kate Witek (state auditor), Scott Moore (secretary of state), and Dave Heineman (state treasurer) all took at least four flights in Sand's plane between July and November 1998.

Of Nebraska's other biggest pork producers, Jim Pillen, who played defensive back for the University of Nebraska Cornhuskers in the mid-1970s, was former coach Tom Osborne's Platte County campaign chairman in 2000. Pillen gave $4,100 to candidates for Congress and president between

1996 and 2000; his partners, Bob and Brett Gottsch, gave nearly $16,000 to candidates in federal races between 1994 and 2000.[15] In the 1998 election, the Gottsches and Pillen gave about $12,000 to candidates in statewide races.[16]

As a candidate for governor, Mike Johanns was aware of the devastating effect of historically low prices on the state's hog farmers. He suggested that one way he could help would be in his selection of a new director for the NDEQ. He told the board of the Nebraska Pork Producers Association that the next NDEQ director would have "a profound impact" on agriculture in Nebraska. He sympathized with hog farmers' impatience with NDEQ's permitting process. Johanns said, "Their response to everything is 'I need more staff,'" and the "problem is there's no one who can pull the trigger over there."[17]

In the best tradition of Nebraska politics, Johanns—who grew up on a dairy farm in Iowa—played the "farm boy" card, expressing sympathy for hog farmers in his first inaugural address. "The historic lows in pork prices have been all but tragic, especially for the small- and medium-sized producer," said Johanns. "As a farm boy, including someone who raised hogs to pay my way through college, I understand the plight facing our farm economy."[18] Later, Johanns told the legislature that state government could have the greatest impact on the agricultural economy "by reducing costly and burdensome regulation" and providing property-tax relief.[19]

Anti-hog-farm activists also sought political answers to their concerns. Hundreds of farmers thronged to the capitol on Farm Lobby Day in 1998 and 1999 to impress the legislature with their numbers and to pull their own senators aside for some personal lobbying. Donna Ziems and Elaine Thoendel often asked Governors Nelson and Johanns to use their authority to enforce NDEQ regulations the way the two women believed they ought to be enforced. The grassroots groups who formed to oppose the trend toward industrialized pork production found willing collaborators in Senators Cap Dierks, Jerry Schmitt, Chris Beutler, and Don Preister.

Senator Stan Schellpeper of Stanton—who by all accounts had tried to find acceptable compromises in the conflict—died suddenly on Easter Sunday 1999, while the legislature was working to amend the Livestock Waste Management Act. To take Schellpeper's place, Governor Johanns quickly appointed Robert Dickey, a big hog farmer from Cedar County.

In 2000 State Senator Jim Jones was anticipating reelection from the Forty-third District. The largest legislative district in Nebraska, the Forty-third sprawls across nearly seventeen thousand square miles of the Sandhills, from Broken Bow to the South Dakota border. Jones was running unchallenged for his third term in the legislature. But some constituents in Rock and Brown Counties were disenchanted with Jones because he opposed increased regulation of livestock—both as a member of the Natural Resources Committee and in legislative debate.[20] Jones had also twice introduced bills to lift some of the restrictions in Initiative 300. Supporters of I-300 encouraged Cleve Trimble, an O'Neill doctor who shared their views, to run against Jones as a write-in candidate.

After Trimble received a surprising 23 percent of the votes in the primary, spending picked up in both campaigns. Trimble—who put about $15,000 of his own money into his campaign—received donations from health-care and labor interests along with nearly $14,000 from the state's largest teachers' union. Contributors to Jones's primary campaign included the Farm Bureau, the Nebraska Bankers Association, the Nebraska Cattlemen, and the Nebraska Chamber of Commerce and Industry. Once it was clear that Jones's reelection was in jeopardy, those groups more than tripled their support. In the end, Jones reported spending $41,756 compared to Trimble's $58,531.[21]

The 2000 Forty-third District race wasn't the most expensive legislative contest in the state that year, but it was one of the closest. On election day, Jones received 52 percent of the votes, with a margin of six hundred out of about fourteen thousand votes cast. With his seat in the legislature safe for another term, Jones said he was baffled by the strength of the opposition. He acknowledged that his criticism of Initiative 300 had generated some of the votes for Trimble, but in talks with the Nebraska Cattlemen, the Nebraska Chamber of Commerce and Industry, and the Nebraska Bankers Association, Jones said he continued to look for ways to change Initiative 300.[22]

In two terms in office, Attorney General Don Stenberg had been under pressure from Friends of the Constitution to enforce I-300. Then in 1999 and 2000 he filed four lawsuits to that effect—all relating to hog operations. Skeptics accused Stenberg of using I-300 enforcement as a way to help his campaign for the U.S. Senate. Stenberg denied that charge and pointed out that he lacked enforcement power until the 1998 legislature passed a bill giving the attorney general subpoena power and staff to do the work.

POLITICS AND SCIENCE

In December 1998 the LB1209 task force rejected any measures that would simultaneously protect the environment and cost livestock producers money.

Critics accused the group of basing its conclusions on politics rather than science. Mohammed Dahab, a university scientist who later became chairman of the UNL civil engineering department, said his role on the task force was "to make sure that science didn't get beat up in the process of discussing political issues. It usually does." Some of that happened, he said, but "in the end, it all worked out fine."[23] But others said that, in particular, the failure to recommend regulating odor was politically motivated.[24]

Avoiding odor regulation may have worked out fine for the legislature, but it had an altogether different impact on county officials who were trying to balance the rights of livestock producers with their neighbors' rights to clean air. In Chase County, residents of Champion Valley were suffering from the odor of forty-eight thousand hogs at two Sun Prairie operations. Citizens packed public hearings to beg the zoning board and the county commissioners not to allow other operations in because the existing hog farms were such a problem.

Like dozens of other county officials statewide, Chase County commissioner Jodi Thompson and her colleagues had to sift through a mountain of documents to sort out how a new operation proposed by Furnas County Farms and Sand Livestock would affect water and air quality in Chase County. Thompson may have been better prepared than most because as a member of the Environmental Quality Council she was familiar with the state's livestock regulations.

Thompson said the legislature's indifference to odor was a way of passing the buck to counties without the protection of the state. She said the legislature's allegiance to local control was a way to avoid responsibility for the environment.

"It's a political issue and they've passed it off," said Thompson. "Yes, we're in control of what we do but they shoved it off on us, and they're leaving it up to us to make some very big decisions."[25] She pointed out that counties have neither the technical expertise nor the money to pay for expertise to figure out how best to protect their air and water from pollution.

Nevertheless, the Chase County Board of Commissioners hired their own expert to review the materials and testimony they received on the Furnas County Farms/Sand Livestock application. The companies sought a permit to build a 2,850-sow farrowing, nursery, and breeding-development operation in the county. Waiting on the county board decision, the Upper Republican NRD also delayed ruling on an allocation of water for the hog farm. Frustrated by the delay, Tim Cumberland, executive vice president of Sand Livestock Systems, wrote state senators complaining that "County officials are being unduly pressured by residents who have been misinformed by out-of-state agitators."[26] He asked the legislature to remove zoning power from the counties and give the NDEQ "the exclusive jurisdiction to regulate livestock in Nebraska. Local county boards should welcome this protection and this will relieve them of a burden many are not qualified to handle."

Tina Kitt, editor of the *Wauneta Breeze*, responded in an editorial saying Cumberland was apparently "having some trouble grasping the concept" of representative government. She also quoted another western Nebraska editorial: "The Nebraska counties . . . don't need the state Legislature to tell them how to manage their land. And they certainly don't need agribusiness-men telling the Legislature to tell them how to manage it."[27]

For nearly a year the Chase County board studied information from its own consultant, from Furnas County Farms and Sand Livestock, and from dozens of people who spoke at public hearings on the permit application. To reduce odor, Furnas County Farms offered to put expensive covers on its lagoons. Nevertheless, in May 2002 the three-member board unanimously rejected the permit. The board considered both economic and environmental factors and concluded "that the designs, plans, management practices, and reliability of the operators of the facility do not afford sufficient safeguards and reliability to protect the general health, safety and welfare."[28] County board chairman Don Weiss Jr. said, "We just tried to take care of our water and our air."[29]

Since 1985 Susan Seacrest, director of the Groundwater Foundation in Lincoln, has spent many years working to educate Americans about how to protect groundwater. Under Seacrest's leadership, the Groundwater Foundation has gained an international reputation for helping communities protect their groundwater. Seacrest says public officials trying to determine whose science to trust also typically apply ethics, values, and common sense to their decisions. "It's never totally a scientific decision," says Seacrest. "The science has to be there, but it has to be something people can live with"—in other words, what's politically doable.[30]

Regarding big livestock operations, what was politically doable in Nebraska differed in proportion to the distance between constituents and policy makers. Nearly every county placed more restrictions on big livestock operations than those the legislature was willing to adopt.

CHECKOFF POLITICS

Pork producers are fond of pointing out that "Pork, the other white meat" is one of the most recognizable advertising slogans in the country. It's a promotion campaign that producers contribute to every time they sell pigs—at the rate of forty-five cents of every hundred dollars of sales. The so-called checkoff raises about fifty million dollars a year. Federal law allows the money to be used only for promotion, education, and research.

After the checkoff became mandatory in the mid-1980s, it was controversial among some hog farmers who considered it a tax. The money is channeled through the USDA to the National Pork Producers Council (NPPC) and to subsidiaries in forty-four states. With rapid consolidation in the industry and thousands of hog farmers leaving the business every year during the 1990s, some farmers complained that the checkoff wasn't doing what it was intended to do—keep them in business.

The NPPC was the focus of much of that discontent because some hog farmers believed it favored big industrial producers over independent farmers. The suspicions of many were confirmed in 1997 when it was revealed that the NPPC had hired a consultant to collect information on groups that opposed industrial and corporate hog production. Targets of the study included Iowa Citizens for Community Improvement, the Missouri Rural Crisis Center, the National Farmers Union, and the Center for Rural Affairs. All had publicly objected to industrialized hog production, and they were all known for promoting family farms and sustainable agriculture. The consultant said these groups weren't to be trusted because they spread misinformation about the modern pork industry.[31] The report caused an uproar because it seemed as though the NPPC had illegally used pork checkoff money to spy on its critics.

In 1998 the Farmers Union and other groups started circulating a petition nationally to do away with the checkoff. They collected more than nineteen thousand signatures from among ninety-eight thousand hog farmers nationwide—enough to put the checkoff up to a vote. In the fall of 2000 over thirty thousand pork producers voted 53 percent to 47 percent for repeal. Secretary of Agriculture Dan Glickman said, "The great underlying theme is the big factory farm versus the small farmer in the pork industry . . . those small farmers defeated the check-off."[32]

Then Anne Veneman—George W. Bush's new secretary of agriculture—invalidated the vote in order to avoid litigation. The decision angered many small pork producers, but it cheered the Nebraska Pork Producers Association. In addition to promoting pork and supporting research, the NPPA used checkoff money to educate willing hog farmers on ways to reduce odor and to protect water from pollution. By the fall of 2001 about forty-five Nebraska pork producers had voluntarily participated in an NPPA program to assess how well they were doing in protecting the environment around their farms.[33]

By October 2002 the future of the checkoff was uncertain, due to the ruling of a federal judge in Michigan. Judge Richard Enslen said the checkoff violated constitutional guarantees of free speech and association by forcing some farmers to pay for activities they opposed. He wrote, "The government has been made tyrannical by forcing men and women to pay for messages they detest. Such a system is at the bottom unconstitutional and rotten."[34] When this book went to press, the USDA and pork producers who supported the checkoff were considering an appeal.

Responding in part to controversy over policies of the National Pork Producers Council, the NPPA had somewhat distanced themselves from the national group. Then in January 2002 NPPA officers ousted executive director Steve Cady, who had held the job for three years. He said his forced resignation was due largely to his sympathies for small producers, but others said Cady had problems in working with members.[35]

In representing all sizes of hog farmers, NPPA leaders have had to strike a delicate balance in making policy because the controversy over mega–hog farms is often cast as a conflict between large and small hog farmers. It's a challenge that confronted politicians in the Nebraska legislature as well.

I'm convinced this would never have been an issue if hogs didn't stink.

– State Senator Ed Schrock, Interview by author, 26 February 2001

Another Pass at the Legislature

When the Nebraska legislature convened in January 1999, senators would become acquainted with a new governor, a new director of the NDEQ, a new chairman of the Natural Resources Committee, and newly riled-up citizens from counties that had recently become home to tens of thousands of hogs. Hog prices had plunged to eight cents a pound—a fifty-year low. While small hog farmers went out of business, big pork producers expanded, increasing neighborhood tension.

Nebraska's situation reflected national events. A national study found that confined animal feeding operations, or CAFOS, were controversial in at least thirty-eight states. In twenty-two states, legislation to regulate CAFOS had been considered within the previous year. Swine generated the most controversy.[1]

In Nebraska, as in other states, both pork producers and their neighbors would look to the legislature for help in solving their problems. The 1999 legislature considered fair-pricing legislation, limits on packer ownership of livestock, and an amendment to the state's ban on corporate farming to help young farmers get started. Six bills would be introduced to amend LB1209—the 1998 law known as the Livestock Waste Management Act.

The 1999 legislature elected Senator Ed Schrock of Elm Creek to replace Lincoln senator Chris Beutler as chairman of the Natural Resources Committee. Beutler considered his loss, in part, a kind of retribution for his efforts to more tightly regulate livestock waste. Others in the legislature said Beutler failed to compromise on the issues and alienated former supporters. Schrock—a farmer and cattleman—called himself a "practical environmentalist." He said, "I want to protect the environment but I want to be reasonable too."[2]

Schrock visited Long Pine and came away thinking there were places that hog confinements could be built without endangering the trout stream. Dave Sands, director of Audubon Nebraska, suggested to Schrock that the legislature should place the watersheds of all such streams—called cold-water, Class A streams—off limits to new livestock operations. There were about fifty such streams in the state, most of them in the northeast and the panhandle. In a bold move for a new committee chairman and a farmer, Schrock introduced LB822—a bill to ban new livestock operations in the watersheds of the state's trout streams.

On the day of the LB822 hearing, while lobbyists for the cattle and hog industries still debated the bill's merits, the state's most prominent fisherman—Tom Osborne—spoke in favor of it. It was Osborne's first public move away from his singular role as the Cornhuskers' beloved former football coach and toward the announcement a year later that he would run for Congress from Nebraska's Third District.

Osborne gave the Natural Resources Committee and a crowded hearing room a short lesson on the trout's delicate relationship to its environment and on the importance of protecting the state's few trout-friendly streams from pollution. He said he didn't want to be interpreted as being antilivestock and added, "It does seem that there are certain areas that are ecologically fitted to raising trout. . . . It seems like there's enough land mass that we can put these type of confinement operations other places."[3]

Many considered Osborne's endorsement of LB822 a major reason for the bill's popularity. While other bills affecting intensive livestock operations had created deep rifts between rural citizens, LB822 was remarkable in that no one opposed it at the hearing. Cattle and pork producers supported the bill after it was amended to allow existing operations to continue and to expand within certain limits.

Once LB822 reached the legislature, it was subjected to about three days of scrutiny and debate. Lincoln senators Ron Raikes and Diana Schimek used the bill to revive emergency zoning. They amended LB822 to allow unzoned counties facing sudden, unexpected developments to adopt zoning regulations from nearby counties for a few months until they could write their own regulations.

RECONSIDERING LB1209

In December 1998 the LB1209 task force had finished its study and reported to the legislature. The nine members of the task force—representing livestock producers, the University of Nebraska, and conservation groups—made recommendations on five contentious issues. First of all, they had studied the possibility of requiring livestock producers to set aside money to clean up their operations in case of closure or spills. Under Nebraska law, landfills, power companies, and manufacturing plants must provide this "financial assurance," but farmers and ranchers objected to similar requirements. They said they couldn't pass the costs along to consumers. The task force decided financial assurance would be too costly for farmers already stressed by low prices for hogs and cattle. The task force also found, somewhat optimistically, that because Nebraska had never yet had to clean up an abandoned livestock operation, there was no need to set up a state-funded account for that purpose. The 1999 legislature would agree on both points, doing nothing to provide for future cleanup of a spill or an abandoned lagoon.

Fees put in place by the 1998 legislature were designed to pay for 25 percent of the NDEQ's cost of regulating livestock operations. Although other industries paid fees that were expected to pick up most of the cost of regulation, the task force opposed placing any further burden on livestock producers for that purpose. The task force concluded, "[T]he public is the ultimate benefactor of natural resource protection via regulatory programs. It is for this reason the Task Force recommends that additional revenues needed for the Title 130 [livestock] program should come from general fund dollars."[4] In other words, taxpayers should pick up the cost of protecting the state from livestock waste pollution. The 1999 legislature agreed.

The task force was more circumspect regarding "carcass liquefaction"—a method only Sand Livestock used to dispose of dead hogs. Sand's method involved putting hog carcasses and afterbirth in deep concrete-lined pits where the material was left to decompose for three years until it liquefied. The liquid was eventually removed and applied as fertilizer to cropground—either through center-pivot irrigation or by knifing it into the ground. Answering critics of this practice, Gary Gausman, president of Sand Livestock Systems, said the practice was equivalent to burial. "Whether we buried the animal whole or buried it in a liquid form, what is the difference?" said Gausman.[5]

The 1998 legislature had responded to learning of this practice by permitting liquefied remains to be disked into cropground but not spread on the surface of the ground.

No other state allowed carcass liquefaction, and no state agency in Nebraska wanted to accept responsibility for regulating it. The law gave county sheriffs the authority to intervene when animal carcasses were abandoned in ditches or around farmsteads. The state's Department of Agriculture had ultimate authority over animal carcasses but had no evidence showing that Sand's disposal method was safe.[6]

Some state officials, senators, and members of the public worried that pathogens would survive in the dead pits and then be spread onto fields, endangering animals and humans alike. The task force said more research needed to be done to protect both human and animal health. For a time, at Sand's request, the legislature considered calling carcass liquefaction "experimental" and letting it continue with oversight from university scientists.

Then in April 1999 Dr. Robert Wills, a University of Nebraska veterinary scientist, announced results of tests he had done on the effluent from some of Sand's dead pits—tests paid for by the company. Wills said he found thirteen different bacteria surviving in the effluent. Although none were common human or swine pathogens, Wills said some were "opportunistic pathogens"—disease-causing agents that could threaten people who were already ill or had weakened immune systems.[7] With this evidence and their general unease about the practice, the 1999 legislature banned carcass liquefaction altogether. Instead, it permitted animals up to three hundred pounds to be composted—a process that had been researched at several universities and found to be a safe, effective way of disposing of livestock carcasses.

To ease the pressure on small producers, the legislature exempted from regulation all farms with fewer than 300 animal units (750 hogs) unless they had previously polluted surface waters with their waste or were likely to.

For nearly two years, hog-farm critics had asked the NDEQ to hold local public hearings on permits for big hog farms. The agency declined, saying that hearings were the responsibility of local zoning boards. As the legislature considered the issue again in 1999, two members of the Natural Resources Committee were central to the debate.

Omaha senator Jon Bruning opposed NDEQ hearings, saying he feared that opponents of big hog farms would use them to harass legitimate businesses. "At my core, I'm a capitalist," said Bruning. "I think if someone wants to risk their capital they ought to be rewarded for that. I don't want them to infringe on other people, but I think the state is healthiest when people are risking capital and the economy's growing."[8] Bruning, some other senators, and livestock lobbyists said local zoning hearings would be the best place for local input.

Hastings senator Ardyce Bohlke, however, supported public hearings by the NDEQ. "The more you act like you're not letting people participate, the more you're likely to make them angry. The industry should have bent over backward to listen," said Bohlke. "We do not give the regular citizens of this state enough credit. When people are given information, a large majority of people in this state make good decisions."[9]

In a compromise, the 1999 legislature gave the public thirty days to send written comments to the NDEQ on proposed livestock operations. It also required the NDEQ to notify counties and natural resources districts when livestock producers applied for permits in their area.

Acknowledging that waste lagoons can leak or rupture, the 1999 legislature banned livestock waste facilities within one hundred feet of any well used primarily for human consumption if the well was owned by someone other than the person with the livestock. Senators Beutler and Schrock persuaded senators to budget $150,000 annually to monitor surface water quality. They pointed out that, at the time, the NDEQ had the resources to test water quality in any given lake, river, or stream only once every five years. Without more frequent monitoring there was no way the agency could know, in a timely manner, whether the water was polluted.

The 1999 legislature once again declined to do anything to regulate odor. The LB1209 task force concluded that odor was "subjective by nature and scientifically difficult to quantify." It said that "Nebraska has unique conditions which call for further research" and that "odor is a complex issue, relating not necessarily to size" but to other things, such as management and climate.[10] The new director of the NDEQ, Mike Linder, said the agency could require a farmer to have an odor-control plan but couldn't enforce it because of the difficulty of measuring odor.[11]

The 1999 legislature accepted these arguments and thus failed to deal with what had become the most immediate problem for a growing number of rural Nebraskans—odor from thousands of hogs.

Hog odor can travel as far as ten miles, but no county was willing to require that large a minimum distance between people and hogs. Few counties had a sparse enough population to make it possible. So neighbors who suffered from hog-farm odor turned to their next available recourse—the courts.

A NEIGHBORHOOD QUARREL

In 1991 Earl and Kathleen Stephens built a retirement home on the Boone County farm where they had lived for forty years. Their pleasant life took an unhappy turn a few years later when two farrowing operations housing ten thousand sows were built near their home—one less than a mile north and another about two miles to the east.

In the summer of 1998 Earl said, "We can't open the windows. We don't dare leave the windows open when we leave because if the wind changes, the zilch or the smell will be in our clothes, in our closet. Our quality of life is nothing like it was before. We're just like prisoners in our house."[12] The Stephenses' son Brad lived a half mile from one of the confinement operations. He sometimes burned feed sacks, filling the air with smoke to cover the rotten-egg stench of hydrogen sulfide from the hog waste.

The Stephens family complained to Jim Pillen, who owned the hog farms in partnerships with several other people, none of whom lived in the neighborhood. When asked about the complaints, Pillen said, "Whenever I've been there, I've never been able to detect odor. Whenever I drive by, I've never been able to detect odor. And whenever there are times that odor would be present, I will guarantee you that there's no more odor than I used to have a few years back out on my father's farm. So I think that I stand firm that the odor from my facilities is not inhibiting people's quality of life."[13] Pillen was planning to expand his operations near the Stephenses' home.

Kathleen Stephens said the odor permeated her clothes dryer. She couldn't hang clothes on her outdoor clothesline. Her grandchildren couldn't play outdoors when the odor was bad. Kathleen told Pillen all this when he once stopped by the Stephenses' home. "He just didn't answer me," she said. "But before he left, he did tell me that I should remember that something good comes from everything. But I don't know what he meant by that."[14]

The odor strained a forty-year friendship between the Stephenses and their neighbor Larry Baker, who sold some of his own farmland to Pillen for one of the hog farms. Baker, who lived next to that farm, said he occasionally noticed the smell. "But to me it's no different than it was whenever I was raising them," said Baker, "so it doesn't bother me any." He recalled how, when he smelled hogs on his own hog farm years before, it was commonly said, "That was money you smelt."[15]

In December 2000 Earl and Kathleen Stephens and sixteen other neighbors sued Jim Pillen and his partners, claiming that four hog operations near their homes made their lives miserable. Pillen's attorney countered by saying the neighbors' own farming operations caused the odor, that the economic benefits of the hog farms to the community outweighed any unpleasant effects on the neighbors, and that at least some of the plaintiffs had "heightened sensitivities and not reasonable, typical or normal ones" and therefore weren't entitled to any compensation for their discomfort.[16]

The neighbors didn't ask for damages in the case; they only wanted relief from the odor. In an October 2002 ruling—nearly two years after the lawsuit was filed—Boone County District Judge Michael Owens gave Pillen and his partners a year to reduce odor enough "to enable the Plaintiffs to regain that which has been taken from them" or to "cease operating" the hog farms.[17]

Fishing was a lot better then when we had the runoff. Course,
we didn't have all the pesticides, insecticides, and fertilizer that
we use now. So I think Mother Nature was meant for . . . for
animal waste to . . . to go into the streams to help fish. That's not
the big problem in the way I look at it. I could be all wrong.
– State Senator Ray Janssen, Debate Transcript,
Legislative Bill 870, 5 May 1999

Building on Sand

About thirty miles southeast of Arthur is a windmill like hundreds of others in the Sandhills—wooden, about thirty feet high, with metal blades. Often such windmills are situated in low meadows, perpetually spinning in the reliable Nebraska wind, feeding water from a pipe into a broad, low metal tank that spills over into the abundant grass. All of these windmills draw upon the Ogallala Aquifer, that massive underground storehouse of water stretching from Texas to South Dakota. Residents of the Nebraska Sandhills are justifiably proud that the portion of the Ogallala that rests beneath the Sandhills is relatively free from contamination.

To this particular Arthur County well and the lush forty-acre meadow that surrounded it, Jim Lawler had, for twenty years, brought his pregnant heifers for summer grazing. The water—pure enough for both people and cattle to drink—could be counted on in the hottest, driest months. But beginning in the spring of 2001, Lawler stopped running his heifers here because the nitrate level in the well had reached nearly twenty-two parts per million—twelve points higher than the EPA safe limit for humans and a level that Lawler's veterinarian said could cause the heifers to abort. Lawler was disappointed but not surprised to learn that the level of nitrate in his water had risen from less than one part per million to twenty-two parts per million (ppm) in a little over two years.[1]

On a hill north of Lawler's well are the long, low barns of a fifty-seven-hundred-hog farrowing operation that opened in the fall of 1999. A project of Enterprise Partners and Sand Livestock, the facility looks out of place in this remote landscape, with sandhills bulging on every horizon.

Before the barns were built and the pigs moved in, Lawler and other nearby ranchers had objected to the way the hog waste would be handled. The Enterprise Partners pigs were expected to produce nearly 3.5 million gallons of manure and urine a year. As with hundreds of other such operations across Nebraska, the waste would periodically be flushed from pits beneath the barns into a big lagoon. The waste and water would be held in the lagoon until it was applied to crops or pasture through center pivots. According to NDEQ records, the company intended to line the lagoon with at least twelve inches of tightly packed clay to allow no more than the state's maximum quarter-inch per day of seepage.[2] At that rate, nearly ten million gallons of waste could legally escape the lagoon each year. Pure sand separates the bottom of the lagoon from the aquifer twenty-six feet below.[3]

U.S. Geological Survey maps show that the groundwater in this area moves from the northwest to the southeast. Jim Lawler's well is three-fourths of a mile southeast of the hog farm. He thinks the nitrate in his water comes from the hog waste in the lagoon. He said as much to the NDEQ and to the Twin Platte Natural Resources District in North Platte. Officials from both agencies looked at his well records and their own and said there were other, more likely sources of contamination.

Lawler's well is shallow, with an old casing, and there are rodent holes near the top. It has been suggested that the nitrate came from cow manure deposited on the surface near the top of the well. To the west is farmground that has been planted to crops for about twenty years. NRD staff say the well is directly downstream from the cropground. They believe that nitrate leaching from commercial fertilizer applied to that ground coincidentally reached the well shortly after hogs arrived in the neighborhood. Even if seepage from the waste lagoon were to reach Lawler's well, the NRD says it would take more than twelve years to get there.[4] Jim Lawler doesn't accept any of these theories.

Lawler, his wife Peggy, and two of their neighbors showed me the well in August 2001. Lawler is a sturdy, sixty-something rancher with bowed legs, a straight back, and a habit of gesturing with a wide sweep of his arm when he wants to emphasize a point. About the nitrate levels in his well, he says, "I've told anyone who will listen." I ask him what he wants. "I just want them to clean up their act and get the damn things out of here."[5] He means the pigs, of course. Lawler and his neighbors are both baffled and angry that the NDEQ allowed the hog farm to be built here.

Scuffing sand with the toes of their boots, the three ranchers tell me how the heavy trucks that carry feed to the hog farm and carry pigs away spoil the trails that pass for sandhills roads. They point out the two pivots that, on this late August afternoon, are spraying waste onto sand covered only by a thin growth of grass and yucca. "Where do they think that stuff is going?" asks rancher Ron Lage. "Right through the sand into the groundwater."[6] NDEQ regulations require livestock waste to be spread at agronomic rates— that is, only at the rate that crops can use the nutrients. This morning, Lage called the NDEQ field office in North Platte to report that Enterprise Partners was emptying lagoon water onto land with no growing crops, but he has received no response. He doesn't expect to; he says the NDEQ often doesn't return his calls.

Some ranchers have admitted to a bias against pigs—especially in concentrated numbers. For generations, cattle have spread out across the Sandhills, spreading their manure with them—in nothing like the concentration of manure stored in the lagoon at the Enterprise Partners farrowing site. The 1997 census of agriculture found no hogs in Arthur County. An Enterprise Partners manager said the absence of hogs made the county an ideal place for this operation because the chances were slim that diseases would enter the herd from other farmers' pigs.[7]

With a per capita income of $10,656, Arthur County is the seventh-poorest county in the United States.[8] With only 444 people, it's also one of the least populated, so another attraction of this site is its isolation. The nearest neighbor is about four miles away—far enough to escape disagreeable odors from the facility if it is well managed.

There is abundant, accessible water at this site—some very near the surface. Less than a mile north of the barns is a lush, green valley where Bucktail Lake lies surrounded by tall grasses and cattails. To the south is a wetland.

In 1998 Arthur County was unzoned, so there would be no restrictions on the hog farm except for those set by the NDEQ. The agency permitted the hog facility to be built because it met all of the NDEQ's design requirements.[9] But those requirements were in dispute from the beginning. In efforts to provide oversight for the design and building of the lagoon, the NDEQ would irritate both Chuck Sand and residents of Arthur County.

In the summer and fall of 1998 the NDEQ was putting into effect the Livestock Waste Management Act passed in April, which included some new requirements for hog farms. Like many other big hog producers in the state, Chuck Sand—a partner in Enterprise Partners and owner of Sand Livestock—was expanding. Sand wanted to begin building the farrowing site in Arthur County.

When Sand became frustrated by having to jump through NDEQ hoops, he faxed Governor Nelson: "Ben, As usual, we seem to be being stonewalled by D.E.Q. It is so damn discouraging it is starting to get me down! Do you have any advice or is there anything you can do??"[10]

Sand's manager at the site, Gale Schafer, was concerned that winter weather would delay construction until spring and that an order for breeding hogs would have to be canceled and construction workers laid off from the eleven-million-dollar project.[11] Governor Nelson forwarded Sand's fax to the NDEQ, but there's no record that Nelson put any other pressure on the agency to speed up the process. On the day of Sand's plea to the governor, the NDEQ sent Enterprise Partners a letter asking for more information, which meant more time would pass before construction could begin.

Meanwhile, the Arthur County board—with its own doubts about the safety of the proposed operation—invited members of the legislature to visit the site. The four senators who came were met by about seventy-five people, most of them opposed to the operation. When Senator Ed Schrock, the new chairman of the legislature's Natural Resources Committee, learned of their concerns for the groundwater, he said, "Let's approach this in a rational manner. I know you are upset, but you need to work with your county officials."[12]

After the senators' visit, Larry Sitzman—who had recently left his position as Ben Nelson's agriculture director to go to work for Sand Livestock—wrote a letter to the *Lincoln Journal Star* criticizing the Arthur County crowd for being close-minded. He pointed out the economic benefits of the facility—fifty thousand dollars in annual property taxes, a payroll of nearly four hundred thousand dollars, and a market for local grain.[13]

The Arthur County board responded by passing a temporary weight limit for traffic on the roads leading to the hog-farm construction site. Enterprise Partners asked the Arthur County district judge to put a restraining order on the weight limit, but the judge declined.

Problems with the clay liner would further delay the project. Earthen hog-waste lagoons are basically pits lined with about a foot of packed clay. The clay is typically applied to the bottom and sides of the pit in two layers, watered, and packed down with heavy equipment. The more clay in the liner, the better the seal. But the liner of the Enterprise Partners site failed permeability tests because too much sand was mixed with the clay. The company rebuilt parts of the lagoon liner and retested it to meet specifications. Even so, an NDEQ engineer who reviewed the plans wrote that he wondered if the standard one foot of clay was sufficient protection since only pure sand separated the lagoon waste from groundwater.[14]

The NDEQ doesn't mandate the thickness of lagoon liners, but the industry standard is twelve inches. Tests of the Enterprise Partners' lagoon showed that several thousand square feet of the finished liner was less than twelve inches thick.[15] The company's tests showed that the liner, nevertheless, met the state's permeability requirements.

So the NDEQ allowed Enterprise Partners to revise its construction plans to indicate the liner would be nine inches thick.[16] The NDEQ approved the new plans with the lagoon already in place. Three days later, NDEQ inspectors visited the site and found that the liner on the sides of the lagoon had eroded—before water was put into it.[17] Studies show that lagoons leak more easily from the sides than from the bottom—especially if they're eroded.

The NDEQ told Enterprise Partners to fix the erosion and, in August, authorized the company to put waste in the lagoon.[18] By October the erosion was worse and the same NDEQ engineer who questioned the reliability of a one-foot liner on sand wrote, "[G]iven the limited thickness of the liner at this facility, liner integrity may be in question and the potential for excessive seepage appears to exist."[19]

Liner erosion had also been discovered at an Enterprise Partners nursery in Perkins County. In addition, the Army Corps of Engineers charged that Sand Livestock and Enterprise Partners had illegally dug clay from a wetland to line a Perkins County hog lagoon. The companies said they believed the land where they dug the clay was a cornfield but nevertheless worked with the Corps of Engineers to restore the wetland.[20] The EPA fined Sand Livestock and Enterprise Partners seventy thousand dollars for disturbing the wetland. But the EPA eventually dropped its complaint when the U.S. Supreme Court ruled in an Illinois case that the EPA and the Army Corps of Engineers had only limited authority over isolated wetlands.[21]

The combination of the federal charge and the eroding lagoons did nothing to increase public confidence in Enterprise Partners or in government oversight.

Neighbors had noticed the erosion at both Enterprise Partners sites and complained to the NDEQ. In a letter to NDEQ director Mike Linder, Ron Lage wrote, "We who live in the Sandhills have tried to impress upon the DEQ the great vulnerability of the soils and the high water table in this area and the folly of siting hog confinement lagoons here."[22]

In November the NDEQ asked the company to determine if the liner was leaking and to propose a way to fix it.[23] Lage wondered why the NDEQ didn't use its own engineers to examine the liner. He said, "They're asking the guys that are going to use it if it will work. Well, of course they're gonna say it's gonna work."[24]

Although citizens may be frustrated by what some have called a "fox watching the chicken house" approach to enforcement, it is standard practice for environmental agencies—including the EPA—to rely on the word of an operator's engineer on design and testing. It's partly a matter of resources. "Obviously you'd like a policeman on every corner as well," says Mike Linder. "But you try to balance what the citizens of the state are willing to provide for resources, and that's determined by our legislature."[25]

The Enterprise Partners site in Arthur County has received a lot of attention from local residents, the NDEQ, and the press. I include it here not only for its notoriety but also because the way the NDEQ dealt with the site's problems is typical of the agency's enforcement methods. Linder calls it "compliance assistance," which he says has, to some degree, replaced previous "command and control" enforcement. "Wherever we can, we move towards mediation—towards an amiable resolution where both sides agree what is best for a given situation," says Linder. "The fact is that we see much greater environmental results when we can avoid the judicial process."[26]

The NDEQ's approach reflects a national trend toward what *Governing Magazine* calls "playing down hard-line legal enforcement and trying to help people figure out how to comply with complicated rules."[27] But for those who fear for the purity of their air and water, this balance between hard-handed enforcement and collaboration is hard to swallow.

For several months the NDEQ and Enterprise Partners negotiated ways to repair the lagoons in Perkins and Arthur Counties. Impatience rose on all sides. The grassroots group Save Our Rural Resources hired a former EPA attorney to advise them. She alerted the media to the erosion. In a letter to the NDEQ, the Arthur County attorney pressed for action on the Enterprise Partners lagoon and sent a copy of her letter to Senator Gerald Matzke, who represented the area in the legislature.[28]

Finally, the NDEQ sent a notice of violation to the company, ordering it to stop putting manure and water in the eroding lagoons in both Arthur and Perkins Counties, remove waste from the lagoons, and fix the liners.[29] Then Sand Livestock's attorney wrote the legislature's Natural Resources Committee saying the company had never put manure in the lagoons and was doing its best to comply with the law.[30] By late spring 2000 Sand installed a plastic liner on the sides of the lagoons. The lagoon bottoms would continue to be protected only by clay liners.

When the Arthur County lagoon was being built, the NDEQ was preparing to limit large livestock lagoons to a seepage rate of no more than one-eighth inch per day, but the new rate didn't take effect in time to apply to the Enterprise Partners site. The lagoon could legally seep at the state's current rate of one-quarter inch a day—amounting to about 6,800 gallons per day for each acre of lagoon surface area. That meant the four-acre lagoon could legally leak at the rate of 27,200 gallons a day.[31]

Lagoon seepage is acknowledged in the livestock industry and among scientists and environmental regulators. Considering the high water table and the ease with which water permeates the sand in the area, the USDA's Natural Resources Conservation Service (NRCS) warned that without careful management both the Enterprise Partners lagoon and the application of waste to nearby cropground could threaten the groundwater. The sandy soils have "severe limitations for use as sewage lagoons due to their rapid permeability," said a NRCS report on the proposed site. And "the depth to the water table ranges from zero to several feet in some critical locations of the proposed waste application areas."[32]

Recognizing the potential for contamination, the NDEQ required monitoring wells around the Enterprise Partners lagoon. Elsewhere in the state, an industry or business required to monitor water quality would collect its own water samples and deliver them to a lab for analysis. In the area covered by the Twin Platte NRD, which includes Arthur County, the NRD staff does the sampling; the company (in this case, Enterprise Partners) pays for the analysis. Twin Platte manager Kent Miller said, "We get to see the results of the samples and they're getting an unbiased collection of samples."[33]

Enterprise Partners installed three monitoring wells around their lagoon; the NRD installed two monitoring wells of its own in fields where the waste was being applied. The NRD samples the wells and reports results twice a year to Enterprise Partners and the NDEQ. Tests show that the nitrate in the three wells around the lagoons ranges from 5 to 13 ppm. By spring 2002 none of the monitoring wells under NRD oversight showed any elevation in nitrate level since monitoring began in 1999.

Jim Lawler also arranged for samples from his well to be analyzed. In

January 1999 his nitrate reading was .05 ppm. Pigs moved into the Enterprise Partners barns in July 1999, and the lagoon was approved for use in August. By April 2000 the nitrate in Lawler's well had risen to 18.54 ppm. By October that year it was 22.12 ppm.[34]

Lawler has considered having expensive tests done on his well water to determine the source of the nitrate. Michael Jess, former eighteen-year director of the Nebraska Department of Water Resources, says Lawler's chances of proving anything are slim. Jess, who is intimately familiar with water law, says the mere appearance of cause and effect doesn't prove anything. "You're not going to carry the day unless you can prove it," says Jess. "To prove it means hiring people with scientific disciplines to do isotope tests and mathematical modeling and test drilling to confirm the existence of the aquifer and the physical coefficients and all that stuff."[35] Employing geologists to evaluate the aquifer and the performance of the wells would be costly and time-consuming.

Everyone involved—including partners and managers for Enterprise Partners and Sand Livestock—has expressed concern for the safety of the groundwater at the Arthur County hog farm. But even with the NDEQ and the NRD monitoring the site, there are no guarantees. "We cannot say there's not going to be a problem from the lagoon and from the application of the waste," says Kent Miller, manager of the Twin Platte NRD. "All we can tell people right now is that with the programs we have put in place and the direction and movement of the groundwater, what may be happening in certain locations."[36]

This uncertainty does little to satisfy critics who want outright restrictions on where big livestock operations can be built. Even a longtime NDEQ employee who wished not to be identified said livestock waste lagoons should probably not be built in the Sandhills because of the vulnerability of the groundwater.

Former NDEQ director Randolph Wood once said, "We are under constant scrutiny to assure that we are not overly aggressive nor underprotective."[37] That pressure is reflected in the way the NDEQ carries out its mission. Internal documents show that the agency has interpreted its authority narrowly, strictly within limits imposed by the legislature. For example, but for specific exceptions in state law, the NDEQ says it has no authority to bar livestock operations from locating anywhere in Nebraska. Those exceptions include the watersheds of trout streams, places where the groundwater is closer to the surface than four feet, and within one hundred feet of a domestic well used by someone other than the livestock producer.

In 2000 the Environmental Quality Council rejected a citizen petition to ban livestock waste lagoons from environmentally sensitive areas of the state like the Sandhills. The proposal was supported by environmental and citizens' groups but strongly opposed by pork producers, the Nebraska Cattlemen, and the Farm Bureau.

Even with the state's ban on big livestock farms near trout streams, the NDEQ allowed a dairy to expand near the headwaters of Verdigre Creek in northeast Nebraska. The operation's size fell within the guidelines set by the legislature. In that case, trout fishermen, citizens' groups, and some senators charged the NDEQ with failing to adhere to the spirit of a law that was meant to protect the most pristine areas of the state from pollution.[38]

It was largely in response to public concern in the late 1990s that the NDEQ began to require monitoring wells around livestock facilities on fragile sites—such as in sandy soil with a high water table. Starting in late 1997, as operations enlarged or changed hands, the NDEQ evaluated the need for monitoring wells and required them in sensitive areas—both as a way to document water quality at each site and to gather more data on the quality of the state's water.[39]

Because of widespread disagreement about what constitutes "sound science" on lagoon seepage, the NDEQ also commissioned a University of Nebraska scientist to do a study. In 1998 Dr. Roy Spalding, director of the UNL Water Sciences Laboratory, put monitoring wells around lagoons at thirteen livestock operations in Nebraska. They represented a range of sizes and ages of operations; one was a dairy, one was a cattle operation, and eleven were hog farms.

Levels of nitrate found beneath the lagoons ranged from undetectable to 267 ppm.[40] The highest nitrate level was found beneath an abandoned swine waste lagoon in Lincoln County that had been filled with soil. Nitrate levels beneath two other operating hog lagoons exceeded the federal standard. All three lagoons were built on sandy soil over shallow water tables.

Spalding used an isotope test to determine that the nitrate came from hog waste and not from commercial fertilizer. A biological reaction beneath three lagoons actually caused nitrate levels in the groundwater to decline.

Many news reports on the study said it showed that most hog lagoons in the state don't leak; Spalding said his study should "allay some of the fears about groundwater contamination and the impact from the lagoons."[41] But there were some who found little comfort in the fact that nearly one-quarter of the lagoons in the study—those over sandy soils—were polluting the groundwater. Spalding and NDEQ scientists admitted that thirteen sites with only three monitoring wells at each site wasn't a large enough sample to provide conclusive results.

No effort was made to identify pathogens in the seepage; Spalding did find antibiotics in the lagoons but not in the groundwater. The discovery of high nitrate levels beneath the abandoned lagoon demonstrated the need for proper cleanup when a hog farm closes. The NDEQ also said the study confirmed the agency's practice of requiring monitoring wells around lagoons built on vulnerable sites. The agency would like to do more studies of lagoons' effects on water quality but has run out of money for that kind of project.[42]

There is a growing body of scientific research on lagoons done in other states, but each study has critics who say it may not apply in Nebraska. For example, a 1998 Iowa study looked for chemicals and pathogens in both groundwater and surface water near earthen hog lagoons. The study was led by the Centers for Disease Control and involved twenty-three scientists from the CDC, the Iowa Department of Public Health, the University of Iowa, and the U.S. Geological Survey. They analyzed samples of surface water and groundwater collected around nine swine operations with more than one thousand confined animals. The samples included liquefied manure from the lagoons, surface water from rivers and drainage ditches, groundwater from lagoon monitoring wells and private wells, and groundwater from fields where manure had been recently applied.

In groundwater and surface water, the scientists found traces of pathogens that were in the lagoons—including salmonella, *E. coli*, cryptosporidium, and enterococcus. All are bacteria that can make people ill. The study also found antibiotics, nitrate, and trace metals in groundwater and surface water around the lagoons.

The scientists concluded, "These discoveries suggest the possibility of the movement of both chemical pollutants and microbial pathogens through soil and away from their point of highest concentration, the animal manure lagoons, and by overland flow away from the site of manure application."[43] None of the samples in the study were taken from an area where pollution could threaten human health. However, the scientists recommended more research "to accurately determine the potential level of risk, possible pathways of exposure, and critical control points to avoid any potential exposure to humans."[44]

A North Carolina study of nearly sixteen hundred private drinking wells near big swine and poultry operations found that one in ten of them had unsafe levels of nitrate. But that study didn't distinguish among contamination sources, which included chemical fertilizer, septic tanks, and livestock operations. Only three cases of nitrate contamination could be traced to hog or poultry farms.[45]

Kansas State University scientists determined that a seepage rate of less than one-quarter inch per day could be achieved if clay-based soils were compacted in a liner at least twelve inches thick. The scientists also concluded that manure on the bottom of the lagoon reduced seepage and that when lagoons were built in coarse-textured soils (such as sand) above high water tables there was "appreciable contamination and seepage." They also found negligible nitrate contamination "when the depth to the water table was greater than 100 to 130 feet."[46] Jay Ham, who led the study, said, "[T]he whole Middle West has good soils for making lagoon liners, but you'll never get zero seepage from an earthen lagoon."[47]

Harmful components of lagoon seepage such as bacteria and nitrogen are thought to be filtered through the soil before reaching groundwater, much as a domestic septic system works. But Rick Koelsch, a UNL professor of biosystems engineering who advises pork and cattle producers on manure management, says more research needs to be done on the filtering properties of soil.

Koelsch says there's no problem if only water is moving through the soil, but if the water is carrying contaminants that's a problem. "At this point in time the science has told us that a fine-textured soil is a very good filter and we can rely upon it to a certain extent," says Koelsch. "But we've got to recognize what are high-risk situations, such as shallow water tables and sandy soils. We've got to take far greater measures in those circumstances. If we're dealing with heavier clay soils we probably can live with that."[48]

To this point—despite the uncertainties—the EPA, the NDEQ, the Nebraska legislature, and policy-making bodies in most other agricultural states have concluded that some groundwater pollution from livestock waste lagoons is acceptable. And although levels of nitrate that exceed 10 ppm can cause health problems for infants and pregnant women, citizens may be somewhat indifferent to it because it's invisible, odorless, and tasteless.

Dennis Schueth manages the Upper Elkhorn NRD, including parts of northern Holt County where nitrate levels exceed the federal standard in 46 percent of monitored wells.[49] "We need to decide what contaminant levels to accept," says Schueth. "We can be more environmentally sound if we want to pay more for our food."[50]

The alternatives to clay liners are all more costly and all have limitations.

The seams in synthetic liners may leak. Concrete may crack. Monitoring wells reveal pollution only after it occurs.

RUNNING OFF

Many who speak for the hog-confinement industry say center-pivot irrigation is the most efficient, economical way to move liquid waste from lagoons to crops. But without careful management, shooting waste through the air contributes to odor problems, and spreading manure on fields without disking it in can contribute to nonpoint source pollution—that is, runoff into surface water.

Hog manure is particularly high in phosphorous, which clings to soil particles that run off in heavy rain or snowmelt. Agricultural runoff containing sediment, fecal coliform and other bacteria, nutrients (nitrogen and phosphorous), and pesticides is the primary cause of pollution in Nebraska's rivers, streams, and lakes. In 2000 the NDEQ assessed sixty-five hundred of Nebraska's approximately sixteen thousand stream miles and determined that 58 percent were impaired. Agricultural nonpoint source pollution was the major cause.[51]

Excess nutrients—primarily phosphorous—contribute to lake eutrophication in Nebraska. Of the state's 514 public lakes, 89 were assessed for the NDEQ's 2000 annual Water Quality Report. Of those, 85 were either eutrophic or hypereutrophic, meaning that excess nutrients in those lakes fed algal growth.[52] When the algae die, they use up oxygen necessary for the survival of other aquatic life, such as fish. Because nonpoint source pollution mingles contaminants from many sources, it is difficult to measure the impact of any one source, such as animal waste. That impact hasn't been sufficiently studied to satisfy all sides in the debate over concentrated feeding of livestock.

In 1995 a consortium of thirty-three scientists, sociologists, economists, and medical experts concluded that "the most pressing need is for local and regional surveys of actual confinement facilities, the performance of their pits and lagoons, their land use practices, waste disposal methods, ammonia emissions and atmospheric deposition, and the chemical, physical and biological integrity of surface and ground waters that they can potentially impact. It is not sufficient to base answers to citizens' questions on theoretical ranges based on state regulations, engineering specifications and waste management plans, because as the saying goes, 'the road to hell is paved with good intentions.'"[53]

Those scientists concluded as well that studies needed to be done regionally because intensive animal agriculture—including dairy, swine, cattle, and poultry—is so widespread as to be affecting regional ecosystems. Nebraska's waters join those from the entire midsection of the country in the Mississippi River and on to the Gulf of Mexico where an excess of nutrients has created a vast dead zone in gulf waters.

The U.S. Environmental Protection Agency has concluded that agricultural runoff contributes to pollution in 60 percent of impaired rivers and streams and 30 percent of impaired lakes in the United States.[54] New EPA regulations issued in December 2002 require thousands of the biggest livestock operations to reduce runoff of both nitrogen and phosphorous. States have two years to comply with the new regulations and will set their own nutrient limits.

Nebraska officials were critical of a draft of the EPA regulations. Agriculture director Merlyn Carlson said they would "work to further degrade states' rights in protecting their individual natural resources." Carlson also questioned whether the EPA has authority to regulate nonpoint source pollution under the Clean Water Act.[55] Governor Ben Nelson opposed the standards "because they amount to a 'one-size-fits-all' approach to a problem that the states can deal with more effectively and efficiently."[56] The NDEQ said states should be allowed to run their own regulatory programs "without overly prescriptive requirements, to allow for the variability in the livestock and crop production industries."[57] Livestock groups were similarly critical of the proposed regulations, but environmental and citizens groups who wanted more limitations on concentrated livestock production said they didn't go far enough.

Anticipating federal regulation, the Nebraska legislature took a somewhat greater interest in water quality. The 2000 legislature asked the NDEQ to study water-quality monitoring statewide to determine whether it is sufficient and scientifically sound. The study concluded that although many groups and agencies do some monitoring of surface water and groundwater quality, there is no coordinated effort to do so.[58]

Thus, despite the best attempts by the NDEQ, the NRDS, county health departments, environmental groups, the Army Corps of Engineers, and the U.S. Geological Survey, it is unclear how well the state is doing in protecting water quality. The NDEQ study concluded it would cost more than $900,000 to set up a comprehensive network and database to monitor water quality and $3.5 million a year to maintain it. The legislature has not allocated money for this project.

The corruption of the atmosphere by the exercise of any trade or by any use of property that impregnates it with noisome stenches has ever been regarded as among the worst class of nuisances. The right to have the air floating over one's premises free from noxious and unnatural impurities is a right as absolute as the right to the soil itself. . . .

– Nebraska Supreme Court, Francisco v. Furry, 1908

The Smell of Money

Driving south of Imperial in Chase County in June 2001, I open the windows to get a whiff of what people are complaining about. Although I'm within a mile of the forty-eight thousand hogs at Champion Valley Enterprises, I smell nothing.

One problem in documenting odor is that it isn't always present, so people's complaints are typically based upon anecdotes rather than data. Many neighbors of hog farms keep diaries of when it smells. One woman told me the effort seemed redundant. Besides, an NDEQ official told her the agency could do nothing based on the diary and that the only way to test for health-threatening levels of hydrogen sulfide—a major component of hog odor that is regulated by the state—was to place monitors around the operation or on her farm. She was told the monitors cost as much as fifteen thousand dollars, but if several people in her neighborhood shared the cost, the equipment would be more affordable.

The NDEQ didn't offer to pay for monitoring the air quality and usually doesn't unless there is a blizzard of complaints. One family's complaints aren't enough.

JANIE MULLANIX

The Mullanix family lives in a big white house on the side of a hill on what local people call "the South Divide"—the south side of the Frenchman River. From her front steps, Janie Mullanix can see the barns of Champion Valley Enterprises. There are two sites—one a mile directly south of her home and the other about a mile to the southwest. There are twelve barns at each site; two thousand pigs to a barn.

Janie Mullanix, her neighbor Shona Heim, and I settle at the dining room table. They tell me what it is like to live near all those pigs.

Mullanix tells it all at once, with no prompting. "Well, if it's strong, it comes in the house—even with the windows shut, it comes in with a strong south wind. It gives my seven-year-old diarrhea if we have it all day and it makes me sick," says Mullanix. "I don't vomit, but I'm nauseous and I have a tremendous headache. At two o'clock one morning last October the wind was real strong from the south and it woke me up and my head hurt. It was pigs. . . . I had the windows shut. We have new storm windows, and we insulated, and we have new siding. When it's strong you can't keep it out. So I knew I'd be up for a while because my head hurt so bad. I decided to call Mike, because Mike told me before they put it in, 'We have the technology, it's not gonna smell.' And I thought OK, you give them the benefit, you know you don't believe it but you give them the benefit. So I woke him up at two o'clock in the morning. He was real nice. I said, 'Mike, your pigs are stinking. It woke me up. I've got a headache it's making me sick.' And he said, 'We've been trying some new stuff.' And I said, 'Well, it's not working.' And I said, 'I just thought you'd want to know.' "[1] She hasn't talked to Mike Bauerle since that night six months ago. When invited to tell their side of the story for this book, the Bauerles declined to be interviewed.

Janie Mullanix has known Mike Bauerle since she was four. "And they're a nice family," she says. "It just makes me sick. I can't believe he would lie. And David, I never did talk to him before. I think David's probably even a nicer guy but I don't know. Now with David, I guess I'll give him more the benefit of the doubt. I think he probably believed what he was told. Mike, I wonder. It's changed my feelings toward Mike. I think he'd do anything for money now. You know, in the country, you don't treat your neighbors like that. I just think it's rotten. I can't imagine that you could be sick to your stomach and have diarrhea and that it's not dangerous to your health."

In the late 1990s Mike Bauerle, his cousins David and Dirk, and Tim and Steve Leibbrandt made a deal with a subsidiary of Bell Farms of North Dakota to bring pigs to Chase County. The Imperial Chamber of Commerce greeted the decision as a logical way to boost the economy.[2] Mike Bauerle said it made sense to feed local corn to local livestock rather than paying to ship corn out of the area.[3]

The Bauerles and the Leibbrandts were respected citizens of the community so no one questioned the decision—except for Shona Heim, who lives on her family homestead between the two hog farms.

Referring to a state law governing the disposal of herbicide containers, Heim says, "I can't throw a Roundup bottle away in the trash but they can put a million point six cubic feet of pig poop and water within half a mile of my home."[4] When she heard about the plans, Heim began reading up on hog confinements—something she knew nothing about. She had recently planted fifty-four hundred trees on her property and intended to turn it into a wildlife preserve. Now Heim worries about groundwater pollution from the massive lagoons. For a while she kept a diary of the odor. "The first year, we had ten days from Memorial Day to the first of August when it didn't smell, on a scale of five, three or above," says Heim. After several calls to the state's Departments of Health and Environmental Quality she gave up on getting any help from the state.

It is hard for citizens to understand why the state can't protect them from odors that make them and their children sick. People who suffer this way become angry because they're at the mercy of neighbors who don't seem to care, and because state agencies with "health" and "environment" in their names offer no relief. "I don't think our legislature cares," says Janie Mullanix. "They're so gung-ho economic development they're giving away what's good about Nebraska."[5]

Transcripts of legislative debate over livestock waste in 1998 and 1999 show that many senators assumed that controlling odor would have unacceptable costs. Hog prices were starting to sag in the spring of 1998, and senators were cautious about doing anything to further jeopardize pork producers even as they expanded. Besides, it's easier to study contentious issues like odor rather than to pass laws to regulate them. The LB1209 task force concluded that "Nebraska has unique conditions which call for further research."[6] It seemed to say that pig odor in Nebraska is different from pig odor anywhere else.

The task force report reinforced Senator Stan Schellpeper's view. He said, "No one really knows how to define odor this year and what is offensive, what we can live with, what we can't live with."[7] Senator Dave Maurstad, who would soon leave the legislature to run for lieutenant governor, said, "This whole issue of odor and good neighbors, all of that, to me is somewhat puzzling. For any of us that have grown up in the rural area, odor is a fact of life."[8] In December 1998 the LB1209 task force recommended more research into odor issues, but the legislature hasn't funded any research.

The state of Nebraska—like many states—continues to make a fine distinction about odor: it is a nuisance rather than a measurable problem. Calling it a nuisance puts it in the category of a leaky faucet, squeaking door, pesky mosquito, or bumpy road—things that ordinary citizens can put up with for a while or fix themselves. But to those who are forced to endure the invasion of their homes by odor, it is more like a crime than a nuisance. It's hard not to find some justification for a mother's rage when the state agency assigned to protect air quality says it can do nothing about the odor that seems to be making her child ill.

WHAT THE NOSE KNOWS

A miasma of as many as forty gases arises from anaerobic waste lagoons.[9] When molecules of those gases reach the human nose, a complex process turns an odorant—say a molecule of hydrogen sulfide—into an odor. Dr. Donald Leopold tells me this by way of introduction to a lesson on the physiology of smell. Dr. Leopold, an expert in treatment of the ear, nose, and throat, is chairman of the Department of Otolaryngology at the University of Nebraska Medical Center. He explains that molecules of the odorant enter the nose only during breathing.

"You can't smell anything if you hold your breath," he says. "Molecules enter the nose and 10 to 15 percent of them go to the top of the nose and contact cilia [tiny hairs] there that carry the receptors which are enabled by genetically occurring proteins."[10] So what we smell is determined in part by our ancestry. Some people can smell ten thousand odorants, but animals can smell many more. Some people are more aware of some odors than other people are.

"The human sense of smell is quite good," says Dr. Leopold. "Sometimes you need only nine molecules to smell something. That's how moms know teenagers have been smoking." He says the chemical odorant stimulates more than one receptor and you need them all to respond if you're going to determine the correct smell. "So when the molecule of banana goes up your nose, it stimulates more than one receptor." The genetically differentiated proteins inside the receptor allow it to function. "It's the machinery that turns chemical information into electrical information," says Dr. Leopold. "But how we determine we're smelling peanut butter and not bananas hasn't been determined."

He uses a musical analogy to explain. "If you play a five-key chord on the piano, it sounds fine. But play only three of those keys and it's dissonant. People may need twenty-six receptors to distinguish a banana smell. If thirteen are dead, they can't do it and the distortion may make the banana smell like garbage."

"Or make pig poop smell like roses?" I ask. He says a distorted sense of smell works the other way, "It tends to make typically pleasant odors smell crummy." Rather than pig poop smelling like roses it would be the other way around. He says people can be exposed so often to an odor that they become desensitized, which explains why some hog farmers say their hogs don't smell when I'm noticing a powerful odor.

Dr. Leopold says the power of suggestion can influence what someone smells. He tells me about a study in Philadelphia where three sets of graduate students were subjected to the same chemical in a closed chamber. "The first group is told it's a new Chanel perfume, and they love it," says Dr. Leopold. "The second group is told it's an extract from food processing. They're indifferent. The third group is told it's an industrial chemical produced by a known polluter, and they think it's awful."

At this point, Dr. Leopold introduces politics into his mini-lecture. "The people who complain about a smell perceive an odor the same way I do, that is, the process of perceiving is the same for each person," he says. "The difference is what they're smelling and the response to it. Into that goes the politics."

To explore this idea, I ask what level of odorant is needed to make people sick. He says, "Nausea is a response to an input. I could tell you a story to make you nauseous. A child on a bus vomits, and people around the child feel nauseous—nausea is a response to a stimulus."

Suddenly feeling a little queasy myself, I tell him about the epidemiological studies I've read on the effects of hog odor on people—studies done with surveys rather than with clinical tests. I ask him about the value of such studies—where they fit in the scientific literature.

He says he admires well-done epidemiological studies, but clinical information is needed to prove what such studies suggest. He tells me my job is to report on what science there is and then point out that politics will determine how policy responds to the science. "Typically, regulations are made before we have the science," he says. "So is the policy scientifically determined? No, it's politically determined. Is that good or bad?"

Dr. Leopold doesn't answer his own question but recalls that in the 1950s and 1960s doctors told the government that smoking causes cancer—an assertion that has never been clinically proven. That is, scientists don't know what happens at the cellular level that enables smoking to cause cancer or heart disease. Nonetheless, tobacco companies have been forced to pay millions in settlements to the states for the expense of treating smokers who had cancer.

Similarly, no one knows why ninety-four-year-old Mabel Bernard or seven-year-old Nicholas Mullanix become ill when they smell hog waste. "It's amazing what we don't know about the sense of smell," says Dr. Leopold. "Clearly outlined studies can be performed that will give guidelines on what levels make people sick. But it will require active politicking for policy makers to agree on a level that makes people sick."

THE LIMITS OF SCIENCE

Scientists have identified over 160 different compounds in the emissions from hog confinement farms.[11] They come from the barns, the lagoons, and the application of waste to cropground. Through ventilation openings, the buildings themselves emit dust that includes feed particles, swine dander, feces, mold, pollen, grain mites, endotoxin, insect parts, and fungal spores. These dust particles may also carry with them bacteria and molecules of gases like hydrogen sulfide and ammonia.[12] Most big hog operations use anaerobic lagoons to "treat" hog waste. Into these deep lagoons, waste from the barns is flushed and the solids settle to the bottom. What's called "lagoon technology" is basically the natural bacteria in soil and feces that work to break down the manure in the absence of oxygen—thus, the lagoons are called "anaerobic."[13]

Among the gases created by anaerobic digestion are hydrogen sulfide, ammonia, methane, carbon dioxide, and carbon monoxide.[14] Two primary odorants from this mix are hydrogen sulfide and ammonia, but there are others as well. The emissions vary with each hog operation—with the feed ration, with management of the waste, with the number of hogs, the season of year, relative humidity, and temperature. This complexity encourages the official reluctance to regulate odor. There are so many things to measure— not the least of all, human health effects from this chemical and organic stew.

First of all, policy makers must determine which of the odorants to measure. Hydrogen sulfide—which smells like rotten eggs—is the target in several states. The connection between exposure to hydrogen sulfide and health effects on hog confinement workers is well established, but its effect on neighbors is less certain. Nebraska restricts hydrogen sulfide emissions by regulating total reduced sulfur (TRS)—a combination of sulfur compounds.

In Nebraska, any source of TRS—for example, a packing plant, a wastewater treatment plant, or a hog farm—must not emit more TRS than an average of ten parts per million in a minute or an average of .10 ppm in thirty minutes.[15] The NDEQ believes these limits will protect human health. Thus, the regulation is called a "health-based standard." The NDEQ has found no violations of the standard at big livestock operations and has never used the regulation to compel a livestock operation to control TRS emissions.

That fact offers little consolation to neighbors who can detect the odor of hydrogen sulfide long before concentrations in the air reach a level that the state says can make people sick.

In 1997, with a mandate from the state legislature, Minnesota began using a health standard of fifty parts per billion for hydrogen sulfide to measure emissions from feedlots and hog confinements. In February 2000 the Minnesota Department of Health concluded that hydrogen sulfide emissions from a big hog operation in western Minnesota should be reduced "for the protection and well-being of human health."[16] The decision was a victory for Julie Jansen. For five years she had complained to state officials that fumes from the nearby hog farm made the children at her daycare sick.[17]

In Iowa it also took years of pressure from neighbors of big hog farms to persuade the state government to act. In 2001 the grassroots group Iowa Citizens for Community Improvement gathered seven thousand signatures on a petition demanding that the state place limits on air pollution from big hog operations.[18] In response, the Iowa Department of Natural Resources commissioned a study by scientists at Iowa State University and the University of Iowa. Although they couldn't agree on how to regulate odor, the scientists found ample evidence to show that emissions from big livestock operations could pose a health risk for neighbors.[19] As a consequence, the Department of Natural Resources proposed standards for limiting emissions of hydrogen sulfide and ammonia that were still under consideration when this book went to press.

North Carolina is among the many other states where policy makers have struggled with whether and how to regulate CAFO emissions. Recognizing the complexity of the issue, a task force of scientists concluded, "The management of odor emissions from animal operations involves a complex set of scientific, economic, social and political issues. . . . These challenges, however, must be addressed if we, as a state, are to sustain the economic viability of the animal agriculture industry. We must undertake an aggressive initiative to address issues of odor nuisance and potential health effects associated with odors."[20]

There is a well-accepted body of research on the health effects of working in hog confinement facilities. But to this point, all the studies of health effects on neighbors have been epidemiological in nature; that is, they're based on surveys. But the survey findings are consistent.

A study in Dakota City, Nebraska, was the first to connect increased trips to hospital emergency rooms with elevated levels of hydrogen sulfide in the air. The federal Agency for Toxic Substances and Disease Registry (ATSDR) showed a direct association between elevated levels of toxic hydrogen sulfide in the air and an increase in respiratory illnesses among children. Trips to emergency rooms at two hospitals increased by 20 to 40 percent when hydrogen sulfide levels were high.[21] A later epidemiological study by the ATSDR in Dakota City found no statistically significant evidence that long-term exposure to low levels of hydrogen sulfide causes neurological disease.[22]

An Iowa study of eighteen people living within two miles of a four-thousand-sow hog farm found "significantly higher rates of . . . toxic or inflammatory effects on the respiratory tract."[23] These symptoms included nausea, weakness, dizziness, headaches, plugged ears, burning eyes, runny nose, and sore throat—symptoms similar to those experienced by confinement workers.

A larger North Carolina study came to similar conclusions. This study included 155 people living near big hog and cattle operations. Those living near hogs reported "increased occurrence of headaches, runny nose, sore throat, excessive coughing, diarrhea and burning eyes."[24] Another study in North Carolina found that people living near big hog confinements experienced mood changes, including tension, depression, anger, and fatigue.[25]

Except for the Dakota City studies, none of this research measured the level or contents of air emissions. None of these studies included clinical measurements such as blood samples, x-rays, or tests of lung functions on the people in the study. All the researchers said more study was needed.

One study concluded that the mixture of emissions could contribute to both odor and irritation, and that odor can be a warning of potential health symptoms.[26] But there would likely be little consolation in knowing that, rather than the odor making you ill, it is irritation from some other factor—a chemical or particulate.

Susanna Von Essen, a professor of pulmonary and critical care medicine at the University of Nebraska Medical Center, says all these studies are a good start, "But we need lots more science before we make laws that drastically limit how people raise hogs in Nebraska."[27] Von Essen grew up in Cuming

County, Nebraska's biggest livestock-producing county. She doesn't believe the emissions from hog operations threaten public health. "It's annoying, like living in the flight path of a large airport, but there's not a lot of evidence that it will harm your health," says Von Essen. She says epidemiological surveys are a good place to start, "and that's where everyone begins when they enter a new field, but you have to move beyond surveys for policy makers to make rules."

Big hog confinements have typically been placed in isolated parts of Nebraska, so only a handful of people are affected by the odor from those hog farms that fail to control it. Often those who say they suffer from the odor have little influence with policy makers. Bill Weida, an economist who works with the GRACE Family Farm Project, says "a mega–hog farm destroys the social fabric of rural communities because 'it pits neighbors against one another, first in the debate over where and if it should be built, then by adversely affecting some and not others. If you could take all the odor generated by such farms and put it in a balloon and release the balloon in a tent the size of the county, then all people would be affected equally. But those who live closest to the complex are hurt much more than others, and that is a recipe for tearing communities in two. It violates the unspoken rule of not doing anything that harms your neighbors.'"[28]

In Chase County the rift in the community widened as neighbors of Champion Valley Enterprises came forward at public hearings to say how the odor from the hog farms had changed their lives. The Bauerles and the Leibbrandts maintained their silence. During a tense and crowded zoning hearing on a permit for a new Furnas County Farms hog farm in Chase County, David Bauerle, a member of the Planning and Zoning Commission, listened silently as residents of Champion told the commission how odor from the Bauerles' hogs had tainted the community.

In an interview after that hearing, commission chairman Charley Colton said, "What is economic growth worth versus what's safe for the neighbors? How can we be fair to everyone? I certainly wouldn't want an operation near me."[29] When it came time to vote, both Colton and Bauerle voted with the majority to approve the permit. Of the three planning commission members who voted "no," one was Shona Heim.

There are numerous ways to manage hog farms to reduce odor. The waste can be aerated or churned, for example—a process that reduces odor in municipal wastewater treatment plants. A layer of straw on top of the lagoon traps odorants. The pigs' diets can be adjusted to reduce odor in the manure.

Vegetable oil sprays used inside confinement barns keep dust down. Some farmers disk manure into cropground rather than spraying it on through center pivots. Synthetic covers can be put over lagoons.

These methods all cost money, and some producers aren't willing to make the investment or the effort—even for the sake of neighborhood harmony. Although the National Pork Producers Council insists that odors from hog farms don't affect human health, the NPPC spends thousands of dollars in checkoff money to educate pork producers in ways to control odor.

One Nebraska pork producer expects a miracle. Brian Mogenson knows that some neighbors of his hog farms in Antelope County suffer terribly from the odor. "We regret there's any smell at all," he says. "There's people working to make lagoons not stink. My theory is if they can put a man on the moon, they can make hog manure not stink. Till then, we do what we can."[30] So three or four times a year Mogenson gives a ham or turkey to all the neighbors who live within a mile of his hog farms. "I guess I try to heal all the wounds I possibly can," says Mogenson, but he has little sympathy for those who complain. "Anything to do with progress is in some people's eyes detrimental. Livestock smell is part of rural Nebraska. If I was building hog sheds in Lincoln, then I'd say people could bitch about my hog sheds."

Tom and Elaine Kimes, who live a half mile from one of Mogenson's fourteen-thousand-hog operations, are unmoved by Mogenson's offering. They give his hams to charity. "I believe in free enterprise, but I believe they infringed on our rights," says Tom.

"I feel helpless and hopeless," says Elaine. "Our children worry about coming home. Some people have encouraged us to sue, but that doesn't seem right." Tom says, "We don't have the resources to do that."[31]

In 2001 the Nebraska Department of Environmental Quality started requiring odor management plans from all of the biggest new hog operations. Eventually, all large existing livestock operations will also be required to do what they can to control odor. But the NDEQ won't check to be sure the odor-control methods are working.

Measuring odor is an inexact science. The NDEQ has the authority to regulate only total reduced sulfur, which includes the odorant hydrogen sulfide. "Clearly we can't put a monitor at every location," says NDEQ director Mike Linder. "It would be prohibitively expensive."[32] In the absence of more involvement by the state in regulating odor, lawsuits will continue to be the only recourse that some Nebraskans have when they long for a breath of fresh air.

Family farms have given the country a great deal in the way of affordable, wholesome food products. We hope their way of life survives. But if it does, it should be because they can adapt to changing conditions—not because they persuade government to shut down the competition.

– Editorial, *Omaha World-Herald*, 17 September 1997

To Make a Silk Purse Out of a Sow's Ear

Wayne Kaup has a mission statement: "To become a large, diversified agricultural company." A vision of energy and effort, Kaup—tanned, muscular, with a closely trimmed beard and sunglasses—is proud to show off his Holt County hog operation. Gesturing at the neat, grassy slopes around the long white barn, he says, "I like the vision of clean, efficient farms. The lack of cleanliness is a farmer mentality. I raise hogs here, but I hate to go into a store in town and have everyone plug their nose."[1]

No need to plug one's nose on this sunny June afternoon, because there is no noticeable odor from the hogs. There are fourteen hundred hogs in the barn—just the beginning for twenty-eight-year-old Kaup and his brother, Kurt. The Kaup family has farmed for many years in Holt County. Wayne Kaup is determined to continue the business by adopting new methods that suit his vision of successful farming.

"I always wanted to buy a marginal piece of land and improve it with hog manure," says Kaup. "I wanted to get the land at a decent price and sell it as top-yielding cropground."

In just six years Kaup says he has increased the organic matter and water-holding capacity of the sandy soil in this quarter section of land by tilling in the solid waste from his hogs. He stores the waste in a concrete-lined pit and draws off the liquid, mixing it with irrigation water to fertilize his crops. The irrigation pivots have been modified with "drop nozzles" that spray waste down onto the corn rather than shooting it into the air. The method preserves water and nitrogen and reduces odor.

The Kaups and two other Holt County farmers are working with the local university extension agent and natural resources districts to demonstrate how livestock waste can be managed without damaging the environment. For pigs, the process starts with a feed additive that reduces by half the amount of phosphorous in the manure. The Kaups test both the soil and manure to be sure they apply only the amount of nutrient the crop needs—reducing fertilizer costs while improving the soil. The process requires recordkeeping and attention to detail, and it results in a satisfactory profit. "I want to show people there's money to be made in farming," says Kaup. "But I don't believe in acting negligent and profiting from it."

While in high school, Kaup started raising pigs with his uncle. "There were twelve or fifteen guys near Stuart then who raised pigs," says Kaup. "Now it's just us and one other." Kaup's formal education ended with high school, but he credits his success to hard work. "That and a good attitude, management, and knowing your numbers," says Kaup. "Hard work and long hours—it don't sound like a pretty picture, but I'd be bored if I were sitting in a chair."

Between 1997 and 2001 more than three thousand Nebraska farmers quit raising hogs. In that same time, the number of hogs raised in Nebraska dropped from 3.5 million to 2.9 million—down 17 percent in five years.[2] Alarmed by the drop in hog numbers and declining farm income, the 2002 Nebraska legislature again briefly had pork on its plate with a proposal to study the livestock industry. Although many livestock producers favored the idea, it died because hog-farm critics feared an attack on local zoning. For the time being, the issues were laid to rest in the legislature.

Pressures on counties continued. In March the Boone County Board of Commissioners rejected some of Jim Pillen's plans to expand his operations there. But Concerned Citizens for the Cedar Valley—a group that formed to protest the project—kept up their guard, expecting Pillen to apply for more permits. The Red Willow and Hayes County Boards of Commissioners pondered applications for permits from partnerships connected with Sand Livestock; the Chase County Board of Commissioners rejected Furnas County Farms' application for a permit there. Platte County—home base for Sand Livestock Systems and Progressive Swine Technologies—began to consider zoning. By early 2002 all but four of Nebraska's ninety-three counties were zoned.

After Premium Farms lost its case against Holt County in the Nebraska Supreme Court in March 2002, local control seemed to prevail in Nebraska, at least for the moment. Just as Sand Livestock—protesting Nebraska's "hodgepodge" of zoning regulations—began moving some of its operations to Iowa, the Iowa legislature restored some control to local governments.

State-sponsored studies of the environmental effects of big hog farms and a petition signed by thousands of Iowans had helped persuade policy makers to stiffen Iowa's livestock waste law. But Nebraska had no comparable studies for the 2002 legislature to consider, nor were thousands of people objecting to the effects of living near big hog farms. Only dozens of Nebraskans had something to complain about. They included people like Mabel Bernard, Emil Dubas, Kathy and Earl Stephens, Tom and Elaine Kimes, and Janie Mullanix's family—none of them with the confidence, financial means, or experience that it apparently took to influence the Nebraska legislature.

While struggling with a budget shortfall of more than two hundred million dollars, the legislature was unlikely to spend money any time soon to gather information on the long-term effects of big hog confinements on either the environment or human health. All large livestock operations will eventually have to have an odor-management program in place, but the NDEQ will not monitor them to see how effective they are.

Nebraska farmers have typically been good stewards of the land, sparing Nebraskans the environmental catastrophes experienced by other states. Of the few livestock producers who have violated environmental law and regulations—either intentionally or accidentally—most have responded well to the NDEQ's policy of seeking voluntary compliance. According to NDEQ news releases, since 1997 the agency has taken legal action only fourteen times against violators who didn't voluntarily comply with the law or whose violation was egregious enough to warrant a lawsuit. Some examples: wastewater from a cattle feedlot and a hog farm flowed into the Blue River near Polk, hog waste applied to a field killed fish in a farm pond near Tarnov in Platte County, a Cedar County feedlot operator intentionally drained cattle waste into Middle Logan Creek, waste from a cattle feedlot in Nance County flowed into the Loup River, and in Washington County hog waste flowed off cropland into the Elkhorn River. Between 1997 and 2002 the NDEQ collected more than fifty-three thousand dollars in penalties from operators like these.

Although negligent livestock producers are the exception in Nebraska and despite the fact that many who speak for the livestock industry say Nebraska places more restrictions on the industry than any other state, current laws may not be enough to protect the state's water from further degradation. For its 2002 water quality report, the NDEQ assessed 5,174 of the state's 16,212 miles of rivers and streams. In 71 percent, pollution exceeded the state's standard for beneficial uses like recreation, aquatic life, agriculture, and drinking water supply—up from 58 percent in 2000. "Natural and agricultural nonpoint sources" were the major source of pollutants like sediment, fecal coliform bacteria, nutrients, and pesticides.[3]

The extent of groundwater contamination beneath the former National Farms site in Holt County is still unknown. No decision has been made about whether or how to clean up the nitrate pollution beneath the abandoned hog lagoon in Lincoln County that was included in the NDEQ's lagoon seepage study. Because the NDEQ now routinely requires monitoring wells around livestock lagoons built on vulnerable sites, the agency may eventually learn more about how these facilities affect the groundwater.

In a hint of things to come, a hog farmer in southeast Nebraska won a 30 percent reduction in property-tax valuation on his house by arguing that its location near his hog farm decreased the value of his home. The Sierra Club began holding community meetings to teach people how to get their property taxes reduced if they live near big livestock operations.[4]

In Polk County, two relatives went to court—one accusing the other of slander in a dispute over a hog farm.[5] Other lawsuits meandered through the legal system. In the spring of 2002 Sand Livestock and the town of Alma briefed the Nebraska Supreme Court on their differences. Oral arguments were scheduled in January 2003. In southwest Nebraska, lawyers were taking depositions from members of Area Citizens for Resources and Environmental Concerns (ACRES) in Sand Livestock's "strategic lawsuit against public participation," or SLAPP, against them.

A federal judge found South Dakota's anticorporate farming law, which was patterned after Nebraska's Initiative 300, to be unconstitutional. The South Dakota attorney general planned an appeal.[6] In North Carolina, the moratorium on new and expanded hog operations was extended until September 2003.

New EPA rules required all states to limit surface-water pollution from livestock waste but did nothing to protect groundwater or air. Congress passed a farm bill with a record amount of funding for conservation and environmental measures. In particular, under the Environmental Quality Incentives Program (EQIP), livestock producers could receive up to $450,000 between 2002 and 2007 to build facilities to handle animal waste. The old farm bill limited EQIP payments to $50,000 for individual producers. By fiscal year 2007 the federal government is expected to spend $1.3 billion on the EQIP program.[7] The reasoning behind this measure echoed arguments from livestock interests who say that because the general public benefits from a clean environment, taxpayers should pay some of the cost to help producers comply with environmental laws. As members of the House and Senate Agriculture Committees, Nebraska congressman Tom Osborne and Nebraska senator Ben Nelson helped write the 2002 farm bill. Both supported the EQIP funding.

People continued to move away from rural Nebraska. Between 1960 and 2000 the state's thirty-four smallest counties lost more than 31 percent of their population. All are rural counties, with economies based on agriculture. It was predicted that the trend would continue as farms grew in size.[8]

In the boom-and-bust cycle of the West, some wondered what would happen when big hog farms shut down and investors looked elsewhere for profits.

The 1999 Rural Poll had revealed a stark difference between what a majority of rural Nebraskans hoped for the future and what they expected. They preferred a future where most farms were owned and operated by families who lived on them, but they expected that large corporate farms would come to dominate the countryside and that most of the small towns and villages would disappear.[9]

Meanwhile, anti–hog farm activists in Nebraska kept busy. Although Sandhills CARE had successfully resisted Premium Farms' plans in Rock County, CARE members became active on behalf of others in similar situations. They produced a "CARE Package"—a booklet giving advice on how to organize against big hog operations.

For some, the all-consuming fight against the concentrated feeding of hogs expanded into other areas related to agriculture. Long after Premium Farms left Rock County, Lynda Buoy continued to search the Internet each day and sent news stories she found there to anyone interested in farming, the environment, and food safety. Loranda Daniels-Buoy began publishing her writing about farming and farm life. Jim Knopik sold much of his Nance County land, began farming on a smaller scale, and went to work part-time in a blacksmith shop in a nearby town so he could devote more time to advocating for a broad range of sustainable farming practices. Annette Dubas of Mid-Nebraska PRIDE became an organizer for Friends of the Constitution. She also testified at a Nebraska hearing of the U.S. House Agriculture Committee—not about the changes in hog farming but about how hard it often was to feed her family while she, as a farmer, was expected to feed the world.

FAIR PORTIONS

The 4-H swine show had attracted me to the 2001 Furnas County Fair, but I was also drawn to the spacious white barnlike building near the show ring where other 4-H projects were displayed. As in my own days as a 4-H'er forty-five years ago, the walls were hung with children's artwork—tempera paintings, photographs, and drawings. There were glass showcases holding handmade jackets, vests, pajamas, and quilts. There were shelves with breads, cakes, and jellies, and in the center of the building a long table with cucumbers, tomatoes, sweet corn, green and yellow beans, onions, muskmelon, potatoes, and squash. White, red, blue, and purple ribbons indicated a judge's decision about the quality of each item.

4-H'ers have kept up with the times by doing projects in computer programming, PowerPoint, and Internet exploration. But county fairs still celebrate the bounty of harvest, reflecting a deep appreciation for families who work the land and the hope that rural youth will continue the traditions and skills that once made farmers prosper.

What has changed is something that, for want of better words, I'll call "the power of processing and distribution." An example is in the Blue Ribbon building where Furnas County 4-H clubs take turns preparing and serving meals to fairgoers. For supper I pay four dollars for a bottle of Aquafina and what the volunteer behind the counter calls a "walking taco." It is assembled in a thin cardboard carrier: a few corn chips on the bottom, ground beef, grated cheese, tomato, lettuce, cucumber, onion, and some bottled salsa.

I ask where the beef comes from. The question generates blank faces as the query passes from one person to the next behind the counter. I also inquire as to the origins of the onions, cucumbers, tomatoes, lettuce, and cheese. After much searching through empty shipping cartons and squinting at small print on plastic bags, the half-dozen 4-H leaders who prepared and served this supper have some answers. Beyond their purchase at Kelly's—the grocery store on the Beaver City square south of the fairgrounds—the origin of the onions, cucumbers, and beef is unknown. The shredded lettuce comes in a big plastic bag from "Tanimura and Antle, Inc."—names not found in the Furnas County phone book. The tomatoes are from San Diego, corn chips from Frito Lay—as if Frito Lay were a location—and the cheese from Illinois.

The next morning I return for a Blue Ribbon breakfast, and from the volunteers on duty then, I discover that the sausage I eat with my pancakes comes from Quick-to-Fix Foods out of Dallas, Texas. The only food items on the menu from a local source are eggs donated by a Furnas County farmer.

In the middle of a city, I wouldn't be surprised at the diverse and distant origins of my food. But the 4-H buildings and livestock pens at the Furnas County Fair displayed many unprocessed versions of the foods on the Blue Ribbon menu. Watching dozens of farm kids and their parents chow down on walking tacos with ingredients from faraway places, I recalled the farmer who told me he wanted his kids to raise pigs for the fair so they'd know where the pork chop they bought in the supermarket came from.

It would be absurd to generalize for the entire rural population of Nebraska or even of Furnas County from a few hours at the Furnas County Fair, but it appears that even some farmers find themselves distanced from the sources of their food. The USDA estimates that the average distance between the place where food is grown and where it is eaten is fifteen hundred miles.[10]

PICKY EATERS

Some consumers are aware of the uncertain consequences involved in much of modern food production. First, there's concern for what the manure from the concentrated feeding of large numbers of hogs, chickens, or cattle does to the quality of air and water. Then there's the limited amount of research that's been done on the implications for human health and environmental sustainability from the spread of genetically modified organisms.

For at least fifteen years, experts in human health have questioned the heavy use of antibiotics in the livestock industry. The Union of Concerned Scientists estimates that 70 percent of the antibiotics used in America go into livestock—most of it to keep healthy animals healthy and to help them grow faster.[11] Evidence is accumulating to show that bacteria are rapidly becoming resistant to antibiotics—which could mean there will soon be fewer medicines like penicillin and tetracycline available to cure human illnesses.

The European Union has banned the use in livestock of antibiotics that are also used by humans. The American Medical Association, the Centers for Disease Control, the Union of Concerned Scientists, the World Health Organization, and the New England Journal of Medicine have called for either a ban or severe limits on subtherapeutic uses of antibiotics in livestock in the United States.

But livestock interests, such as the Animal Health Institute, dairy, beef, pork, and poultry producers, and even the USDA question the research. They argue that antibiotics are necessary to a viable livestock industry. They say the link between antibiotic use in livestock and microbial resistance to antibiotics outside hog barns, cattle feedyards, and poultry houses hasn't been proven.

FROM FARM TO FORK

While the scientific debate rages on, some consumers are arranging to buy their meat, milk, eggs, vegetables, and fruit straight from farmers who grow it—so they know how it was raised. The USDA says that more than twenty-eight hundred farmers' markets operate in the United States.[12] Their popularity is one sign of a new enthusiasm for "putting a face" on one's food.

Bill Niman wants to help. Niman, a California entrepreneur, is collaborating with some of America's finest chefs and more than 150 hog farmers in seven midwestern states. Niman encourages consumers to select their pork not only for its eating quality but for where and how it is produced, who produces it, and how that production affects the natural and social environment. The center of the Niman Pork operation is Paul Willis's farm near Thornton in north-central Iowa.

Iowa still leads the nation in hog production—more than fourteen million per year—but the number of Iowa hog farmers dropped from 31,790 in 1992 to 17,243 in 1997.[13] Drivers on Interstate 80—Iowa's most heavily traveled east-west route—won't see many pigs. To do that, you have to leave the interstate and drive north on Highway 69 through Wright County. There among corn and soybean fields, the long, low barns of Iowa Select—the nation's sixth-largest pork producer—dominate the countryside. In Wright County alone there are nearly 360,000 hogs.[14] But noticeably absent along Route 69 are the farmsteads that once occupied almost every quarter section—most of them diversified farms where hogs were only one of numerous income sources.

Paul Willis has stayed in business and thrived by using what many would dismiss as an old-fashioned, hopelessly inefficient way of raising pigs. Contradicting the gospel preached at Iowa State University and numerous other land-grant universities in the United States, Willis raises his hogs on pasture open to Iowa's variable weather.

In early September 2001 when I visit, along with a couple of people from North Carolina and California, rain has been drizzling steadily for about a day. Willis—dressed in overalls, a white t-shirt, and heavy leather work boots—gives us each a pair of clear disposable plastic boots to protect our city shoes so we can walk in the hog pasture with him. The slippery grass is a lush mixture of alfalfa, timothy, clover, and fescue. We're warned to avoid touching the thin ankle-high wire that runs through it. The electrified wire is the only method Willis uses to confine his pigs.

Willis says there are about one hundred sows in fourteen pens in this pasture. Several are foraging in the grass, with piglets scampering around them, rolling over each other, rooting in the earth, inspecting their surroundings. They seem to be playing. One grazing sow raises her head, languidly acknowledging visitors as she munches on grass. Someone asks if the sows are dangerous. Willis says we shouldn't worry. "When they're born into this system, they're calm, they understand it," he says.[15]

Across this twenty-acre pasture are scattered small, round-roofed corrugated metal huts just big enough for one sow and her brood. There are also a couple of larger huts for communal gatherings of the pigs. Sows that haven't yet farrowed have each chosen huts of their own and are fussily arranging cornshuck bedding to suit themselves—"making nests," says Willis. He says he has few birthing problems because the sows get so much exercise. In some of the huts, sows lie on their sides in the bedding, grunting contentedly as their families—ten or twelve piglets per litter—scramble over each other to suckle. One sow opens dreamy eyes and snorts at Willis as he leans down to count her piglets.

From nearby there's an angry juvenile squeal; Willis says a sow probably stepped on a piglet's foot, something unlikely to happen in confinement farrowing because there's so little room for a sow even to stand. We ask if he isn't worried about sows rolling over on the piglets. Willis shrugs and says, "I expect a few mortalities, but these are good mothers and their instinct is to avoid hurting their offspring."

When the piglets are five or six weeks old, they're weaned by removing the sows and leaving the young to fend for themselves in the pasture. After up to ten weeks on pasture, the pigs are moved to open sheds at a farmstead to grow to slaughter weight. There they roam freely in a dirt lot.

Within three to five days after weaning, the sows are again receptive to breeding in what Willis calls the "old-fashioned way." No artificial insemination is used on his farm. He says, "What we're about is to allow the pig to behave as naturally as possible."

Willis's operation keeps him and two full-time employees busy. Physical effort is necessary—some heavy lifting, walking through mud, chasing pigs when they get out. "Sure there's some work involved," says Willis, "but it takes two men only two hours to move the huts and set them up." In the spring the hog shelters—called "port-a-huts"—will be moved to a new location so the pig pasture can be planted to soybeans. The next year it will be corn, then hay, which will be harvested and then allowed to grow a second year for the pastured pigs. Iowa's topsoil is famously rich—even more so when naturally fertilized with randomly scattered pig droppings. Pigs aren't just feeding and defecating here but rooting in the moist earth. Willis, who wears a perpetual little smile and has smile crinkles around his eyes, says, "The pigs do the rototilling."

There's a faint odor of ammonia in this field, nothing like the overpowering stench I've experienced at some big confinement operations. "Earth and sun and rain neutralize the manure," says Willis. There's little or no runoff and little threat to groundwater or surface water. The Iowa Department of Natural Resources places Willis's hog operation in a "dry manure" category, meaning he requires no state permit to operate. Willis needs neither expensive engineering nor the capital required to build confinement barns and waste lagoons. He's had pigs on this land for many years, and tests the soil for nutrients before applying manure from the buildings and pens where he keeps hogs he's fattening for slaughter.

Pigs are an important part of a natural cycle on this farm. Their manure provides fertilizer for crops they'll eat. The pigs also dispose of a good deal of household garbage, such as potato peelings from the Willis kitchen—an important function of pigs through the centuries, one that's been abandoned with the industrialization of hog-raising.

Most of the pigs in this pasture are coated with mud; confinement pigs never see dirt, let alone mud. "It isn't that pigs like to be dirty," says Willis. "Mud helps them regulate their body temperature." The extra measure of fat they carry also helps them survive in changeable Iowa weather.

Willis calls his pigs a "composite" breed. Through the mud, it's clear that some have spots, some are black, others are red, and some are pink. They are what is known in swine-raising circles as "Farmers' Hybrid," an old breed of hogs that was once raised everywhere in Iowa. The fat they carry not only helps them thrive in the outdoors but also produces fatter meat—which has fallen out of favor with the American consumer.

Bill Niman is determined to change that preference. He says, "Pigs are all about backfat. There's a balance between backfat, eating quality, and ecology. If hogs are too lean, they won't survive outdoors."[16] Niman has cultivated a network of chefs who also believe that a good measure of fat makes the best pork.

Lean pork won't be found on the menu of Café Rouge, Marsha McBride's upscale restaurant in Berkeley, California. She says her customers like the succulence and juiciness of Willis's pork. McBride is among the chefs at several notable American restaurants who buy their pork from Niman, largely for the eating quality but also because of the way it's raised. "These are happy pigs," says McBride. "And happy pigs make for happy customers."[17] McBride has seen for herself on visits to Willis's farm.

Each year, Bill Niman brings to Paul Willis's farm several chefs who use Niman pork in their restaurants. Gathered there in September 2001 were Michael "Cal" Peternell, Russell Moore, and Lory Podraza of Chez Panisse in Berkeley; Steve Johnson of the Blue Room in Cambridge, Massachusetts; Michael Romano and Ken Callaghan of Union Square Café in New York; Rob Chalmers of Lucques in Los Angeles; Ari Weinzweig of Zingerman's in Ann Arbor; and from Des Moines, Jeremy Morrow of Bistro 43 and Gary Hines of Bistro Montage. Kurt Friese came from Devotay in Iowa City.

"They get to touch a pig and see a pig behave like a pig," says Willis. "I think it's a similar experience for me as a farmer to go to one of these wonderful restaurants and have the experience of tasting something grown on my farm as prepared by a food artist."[18] Willis has become a minor celebrity in Berkeley, where both Café Rouge and Chez Panisse have hosted dinners to honor him and his pork.

Bill Niman says it helps to have begun his enterprise in California, where he says people care about food. Cal Peternell says his customers "always ask about the fish, where it comes from. But I'm not sure they'd ask about the meats unless you tip them off."[19] Restaurants that buy Niman pork, beef, and lamb typically indicate their sources in their menus. It's a practice introduced in Berkeley in 1971 by Alice Waters at Chez Panisse. Waters identifies not only the source of the meat but that of all the vegetables, cheese, fruits, and condiments. Chez Panisse, which has been called "the most influential American restaurant of the past generation," is famous for "farmhouse cooking with grade-A ingredients."[20]

Each year after the chefs visit the Willis farm, they prepare a four-course dinner for Niman's hog farmers. In 2001 the event was held in historic Hotel Fort Des Moines, with white linen and roses on the tables and the food prepared by chefs who, the day before, had walked through Willis's hog pasture. The pork on the menu made its first appearance on an appetizer plate as blood sausage and liverwurst. The main course was pork shoulder braised with fennel, sweet peppers, sage, and potatoes. There were after-dinner speeches by Bill Niman, a representative from the Animal Welfare Institute, and other luminaries, praising the farmers for their husbandry and the quality of their pork. There were awards for best carcass of the past year, best marbling, and best pork by a new producer. "It's a tremendous experience," said Bob Spenner, one of about a dozen Nebraska farmers who sell pigs to Niman. "As farmers we feel appreciated for a while."[21]

Bill Niman said, "The underlying goal is to show the farmers how great their pork is and to make the connection between their husbandry and the end market." Of the chefs who cooked the meal, he said, "These culinary monsters from around the U.S. are important customers and proof of what we do."[22]

The dozen or so hog farmers I met at the Niman dinner all said they joined the operation because they were already raising pork using many of the husbandry practices Niman requires—those recommended by the Animal Welfare Institute. No antibiotics or growth hormones are used, and the pigs are raised on pasture bedding and dirt. No electric prods are used to control them.

Niman protects the farmers from fluctuations in the market by maintaining a floor of about thirty-one cents a pound for hogs and guaranteeing farmers six cents above the market rate. When hog prices dropped to eight cents a pound in 1998, Niman paid his farmers forty-three cents a pound for their hogs. The farmers also own shares in Niman Ranch Pork. The demand for the pork is growing so quickly that Bill Niman hopes to double the number of farmers selling him hogs.

In Nebraska, farmers are also catching on to the idea of connecting with consumers. Farmers Choice, North Star Neighbors, Pawnee Pride Meats, Diamond B Beef, and Small Farms Cooperative all sell farm-raised, locally processed meats to local or regional customers. Not all of these farmers raise their pork on pasture, but they're learning how to adjust to what customers want. Farmers and customers meet at about a dozen farmers' markets across the state.

But without some way to expand beyond the local market, few will make a living selling meat this way—it's too labor intensive with too little payoff. Max Waldo of Waldo Farms says, "There's not that many people in the country or world that are that interested in subsistence-type food production. Even if they had that opportunity there's not that many farmers who have an interest anymore in that small enterprise."[23]

One option is a cooperative of the type that nearly two hundred Nebraska hog farmers have formed to keep control over their product and take some of the risk out of hog farming. Family Quality Pork Processors broke ground for a small hog-processing plant near Petersburg in November 2001. Stan

Rosendahl, a former president of the Nebraska Pork Producers Association, said, "What we plan on doing is going to the consumer or customer who wants the meat and asking, 'How do you want it raised? What do you want us to do with it?' If they want it antibiotic-free, we have producers who aren't using antibiotics. If that market grows we've told our producers there has to be an economic advantage to doing that."[24]

The operation will be small enough to trace the meat from the farmer to the meatcase where the customer buys it. Local grocery stores and restaurants have agreed to sell the meat, which is likely to be more expensive than the mass-processed product that comes from big packing plants like those run by Farmland and IBP. Rosendahl thinks that what he calls "the story" will encourage consumers to choose the co-op's meat over other labels.

"The story is that the source of the meat is family farmers," said Rosendahl. "And we can say we have a higher-quality meat than the commodity product." The meat from Family Quality Pork Processors will be sold as fresh meat. "We want to get the product to the stores within a couple of days of slaughter," said Rosendahl, "so it'll be fresh, not frozen."

Rosendahl's operation and that of Niman Pork will not feed the world. Worldwide meat production grew fivefold between 1950 and 1997, from 44 million tons to 211 million tons. Per capita meat production is now about eighty pounds—twice the 1950 level.[25] Pork and poultry accounted for 76 percent of the increase in meat consumption in developing countries between 1984 and 1998, with China leading the way.[26]

With this prospect for growth in demand for pork, those who would like all hogs to be raised as they are on the Willis farm in Iowa will be disappointed. The spread of large confinement operations is likely to continue—if not in Nebraska and not in the United States, then in places like Brazil and China where the demand for pork is the greatest.

The question remains: Can hogs be raised by the thousands in confinement without destroying the natural resources upon which they depend— like the dwellings repeatedly burned in Charles Lamb's famous essay "A Dissertation upon Roast Pig"?

Whether economic forces will overwhelm the concern for the air and water and the quality of life in rural communities is at least in part in the hands of people who eat pork and those whom they elect to do their will.

Notes

INTRODUCTION

1. Neal Peirce and Curtis Johnson, "Dollars and Scents," *Norfolk Daily News*, 9 July 1998, 4.

2. Melanie Sill, "Needed: Long-Haul Commitment," *Nieman Reports* (Nieman Foundation for Journalism, Harvard University) 53/54, nos. 4/1 (winter 1999/2000): 75–78.

3. James C. Olson and Ronald C. Naugle, *History of Nebraska* (Lincoln: University of Nebraska Press, 1997), 201.

4. Patricia Nelson Limerick, *The Legacy of Conquest: The Unbroken Past of the American West* (New York: W. W. Norton, 1987), 27.

5. Editorial, "The Battle over Hog Factories," *New York Times*, 8 July 1998.

1. IF ECONOMICS RULE

1. U.S. Department of Agriculture, National Agricultural Statistics Service (hereafter USDA NASS), *1997 Census of Agriculture* (Washington DC: U.S. Department of Agriculture, March 1999).

2. The number is calculated from the permitting files for Champion Valley Enterprises and Tim and Steve Leibbrandt at the Nebraska Department of Environmental Quality, Lincoln, Nebr.

3. Mabel Bernard, interview by author, Enders, Nebr., 26 June 2001. Mrs. Bernard died in January 2003, while this book was in production.

4. Joyce Bernard, interview by author, Enders, Nebr., 26 June 2001.

5. Betsy Freese, "Pork Powerhouses," *Successful Farming*, October 1997, 23; October 1998, 21.

6. Jim Pillen, interview by author, May 1998, broadcast in Carolyn Johnsen, "Factory Hog Farms Defined," *Nebraska Nightly*, Nebraska Public Radio Network (NPRN), 3 August and 30 November 1998.

7. Carolyn Johnsen, "Hog Farm Stinks," *Nebraska Nightly*, NPRN, 2 December 1998.

8. Gerald Bodman, telephone interview by author, 2 April 2001. Bodman is a retired UNL professor of agricultural engineering who designs livestock waste lagoons and who has often written on the topic. His comparison of hog and human waste volumes lies in the mid range of estimates made by others in the field.

9. U.S. Department of Agriculture/Environmental Protection Agency, "Draft Unified AFO Strategy, Executive Summary," release no. 0373.98, Washington DC, 17 September 1998, 3.

10. Nebraska Department of Environmental Quality (NDEQ), "Lagoon Seepage Rates in 11 States," 22 March 1999, Lincoln, Nebr.

11. NDEQ, "Lagoon Seepage Rates."

12. Jay Wolf, testimony for Nebraska Cattlemen, Hearing Transcript, LR123 Interim Study (95th Nebr. Legis., 1st sess., Agriculture and Natural Resources Committees), 31 October 1997, Exhibit 11.

13. Wolbach Foods permitting file, Form WP-42, NDEQ Livestock Program, Lincoln, Nebr.

14. Darin Uhlir, interview by author, Wolbach, Nebr., 25 July 2000.

15. USDA NASS, *Agricultural Statistics Database, http://www.usda.gov/nass/pubs/histdata.htm*, retrieved 2 March 2002.

16. Mark Drabenstott, "This Little Piggy Went to Market: Will the New Pork Industry Call the Heartland Home?" *Economic Review* (Federal Reserve Bank), 3d quarter 1998, 82.

17. Kevin Wetovick, interview by author, Fullerton, Nebr., 25 July 2000.

18. Ron Dubas, interview by author, Fullerton, Nebr., 25 July 2000.

19. Annette Dubas, remarks at public meeting in Madrid, Nebr., 28 May 1998, and broadcast in Johnsen, "Factory Hog Farms Defined," 3 August and 30 November 1998.

20. Syd Burkey, remarks to Natural Resources Committee of the Nebraska Legislature, Crete, Nebr., 29 October 1997, and broadcast in Carolyn Johnsen, "Hog Farm Hearing," *Nebraska Nightly*, NPRN, 30 October 1997.

21. Rick Koelsch, remarks to Natural Resources Committee, "Hog Farm Hearing."

22. Drabenstott, "This Little Piggy," 95.

2. HOG-WILD TO EXPAND

1. USDA NASS, *Agricultural Statistics Database.*

2. Randolph Wood, Hearing Transcript, LR123 Interim Study, 16 September 1997 (Fullerton, Nebr.), 3.

3. Julie Anderson, "Agency Catches Up on Hog Lots," *Omaha World-Herald,* 2 November 1998, 21, 23.

4. Paul Hammel, "Emotions Run Raw on Issue," *Omaha World-Herald,* 30 December 1998, 6.

5. "Central Nebraskans Protest Mega Hog Farms," *Grand Island Independent,* 17 July 1997, 1A.

6. USDA NASS, *1997 Census of Agriculture.*

7. Joby Warrick and Pat Stith, "Boss Hog," *Raleigh News & Observer,* 19, 21, 22, 25, 26 February 1997.

8. Lynn Bonner, "Critics Say State Must Do More to Protect Rivers," *Raleigh News & Observer,* 17 August 1995.

9. Minority Staff, U.S. Senate Committee on Agriculture, Nutrition, and Forestry, "Animal Waste Pollution in America: An Emerging National Problem," 105th Cong., 1st sess., December 1997, p. 3.

10. North Carolina General Assembly, "The Swine Farm Siting Act," Senate Bill 1080 (Raleigh, 1995 Regular Session); "An Act to Implement Recommendations of the Blue Ribbon Study Commission on Agricultural Waste," Senate Bill 1217 (1996 Regular Session), "The Clean Water Responsibility and Environmentally Sound Policy Act," House Bill 515, (1997 Regular Session).

11. U.S. Senate, "Animal Waste Pollution," 3.

12. Consent Decree in *Natural Resources Defense Council, Inc. v. Browner,* civ. no. 89-2980 (U.S. District Court for the District of Columbia, 1992).

13. General Accounting Office, "Animal Agriculture: Information on Waste Management and Water Quality Issues," Briefing Report to the U.S. Senate Committee on Agriculture, Nutrition, and Forestry, 104th Cong., 1st sess., GAO/RCED-95-200BR, June 1995.

14. Christopher L. Delgado, Mark W. Rosegrant, and Siet Meijer, "Livestock to 2020: The Revolution Continues" (paper presented at the annual meeting of the International Agricultural Trade Research Consortium, Auckland, New Zealand, 11 January 2001). The authors do research at the International Food Policy Research Institute in Washington DC.

15. Dan Hodges, "Federal Council Seeks Producer Input to Help Solve Low Price Problems," *Nebraska Pork Talk,* Nebraska Pork Producers Association, December 1998, 8.

16. Center for Rural Affairs, *Spotlight on Pork* (Walthill, Nebr., spring 1994 and June 1995).

17. Roy Frederick, interview by author, Lincoln, Nebr., 13 February 2001.

18. David Hendee, "Big Hog Operations Bank on Efficiency," *Omaha World-Herald*, 7 April 1996, p. 6E.

19. Frederick, interview.

20. John Ikerd, remarks at the Harold F. Breimyer Agricultural Policy Seminar, University of Missouri, Columbia, Mo., 16–17 November 1995.

3. RILED UP

1. Lawrence Klassen, telephone interview by author, 3 April 2001.

2. Dan Willets, telephone interview by author, 23 March 2001.

3. Tony Lesiak, telephone interview by author, 22 March 2001.

4. Willets, interview.

5. Emil Dubas, interview by author, Fullerton, Nebr., 13 March 2001.

6. Kelly J. Donham, "Respiratory Disease Hazards to Workers in Livestock and Poultry Confinement Structures," *Seminars in Respiratory Medicine* 14, no. 1 (January 1993): 49–59.

7. Kendall Thu, K. Donham, R. Ziengenhorn, S. Reynolds, P. S. Thorne, P. Subramanian, P. Whitten, and J. Stookesberry, "A Control Study of the Physical and Mental Health of Residents Living Near a Large-scale Swine Operation," *Journal of Agricultural Safety and Health* 3, no. 1 (1997): 13–26.

8. National Pork Producers Council, "Frequently Asked Questions," *http://www.PorkEnvironment.org*, Des Moines, Iowa, retrieved 31 August 2002.

9. Jim Knopik, interview by author, Fullerton, Nebr., 7 February 2001.

10. W. Morgan Morrow, "The Disposal of Dead Pigs: A Review," *Swine Health and Production* 1, no. 3 (May 1993): 7–13.

11. Annette Dubas, interview by author, Fullerton, Nebr., 7 February 2001.

12. Emil Dubas, interview.

13. Thomas Jefferson, as cited in Limerick, *Legacy of Conquest*, 58.

14. Freese, "Pork Powerhouses 1997."

15. "Farmers Take Megahog Fight to Lincoln," *Nance County Journal*, 23 July 1997, 1.

16. Lesiak, interview.

17. USDA NASS, *1997 Census of Agriculture*, and Census Bureau, *U.S. Population Census*, *http://www.census.gov/population/estimates/county/co-99-1*, retrieved 20 January 2002.

18. "It's We . . . Not Them and Us," *Nance County Journal*, 25 June 1997, 3.

19. Lesiak, interview.

20. Willets, interview.

21. Emil Dubas, interview.

22. Greg Wees, "Holt County to Form Committee to Set Zoning Rules, Regulations," *Norfolk Daily News*, 25 July 1997, 2.

23. "Meeting Set On Proposed Hog Farms," reprinted from *Omaha World-Herald* in *Nance County Journal*, 13 August 1997, 1.

24. Bill Hord, "Hog-Lot Moratorium Not in Cards," *Omaha World-Herald*, 17 July 1997, 13.

25. Chris Beutler, interview by author, Lincoln, Nebr., 18 February 2001.

26. Art Hovey, "Politicians Give a Listen to Hog Fight," *Lincoln Journal Star*, 16 August 1997, 1A.

27. Hovey, "Politicians Give a Listen," 6A.

28. Ardyce Bohlke, "Interim Study to Examine Ways to Encourage Development Considering Environmental Concerns," Legislative Resolution 123, 95th Nebr. Legis., 1st sess., 1997.

29. Randolph Wood, LR123 Hearing Transcript, 1–39.

30. Jim Pillen, LR123 Hearing Transcript, 16 September 1997, 51–52.

31. Environmental Defense, "Pollution Locator: Animal Waste," *Environmental Defense Scorecard*, New York, *http://www.scorecard.org*, retrieved 4 March 2002. Environmental Defense said it used the *1997 Census of Agriculture* and the 1997 *North Carolina Agricultural Chemicals Manual* to calculate the amount of livestock waste generated in each state. According to Environmental Defense, Nebraska cattle produced nearly seven times more waste than hogs in 1997.

32. Lowell Stone, LR123 Hearing Transcript, 16 September 1997, 147–51.

33. Susan Arp, written testimony, LR123 Hearing Transcript, 29 October 1997 (Crete, Nebr.), Exhibit 10.

34. Tim Cumberland, written testimony, LR123 Hearing Transcript, 16 September 1997, Exhibit 2.

35. Jeff Lindgren, LR123 Hearing Transcript, 16 September 1997, 166.

36. Pillen, LR123 Hearing Transcript, 44.

37. Ron Schooley, LR123 Hearing Transcript, 16 September 1997, 100.

38. Annette Dubas, LR123 Hearing Transcript, 16 September 1997, 81.

39. *State of Nebraska v. Richard Wells*, 257 Neb. 332 (1999).

40. Patrick Rice, "Notice of Violation," sent to Brian Mogenson, 17 December 1997, Premium Farms permitting file, NDEQ Livestock Program.

41. Johnsen, "Hog Farm Hearing."

42. Carolyn Johnsen, "Livestock Waste Panel," *Nebraska Nightly*, NPRN, 10 December 1997.

43. Johnsen, "Livestock Waste Panel."

44. "A Statement from the Board of Directors of the National Catholic Rural Life Conference," Des Moines, Iowa, 18 December 1997; Carolyn Johnsen, "Hog Carcass Disposal Questioned," *Nebraska Nightly*, NPRN, 25 November 1997.

4. HOME RULE

1. Nebraska Department of Roads and Irrigation, *Twenty-first Biennial Report of the Department of Roads and Irrigation*, 1935–1936, Lincoln, Nebr., 57–59.

2. James Denney, "The Place Once Known as Death Valley," *Sunday (Omaha) World-Herald Magazine of the Midlands*, 25 May 1975, 17.

3. Ann Brown, "Proposed Hog Operation Sparks Concern," *Kearney Hub*, 26 July 1997.

4. Carolyn Johnsen, "Local Governments Respond to Hog Farms," *Nebraska Nightly*, NPRN, 7 December 1998. Gary Gausman of Sand Livestock provided the figures.

5. USDA NASS, *1997 Census of Agriculture*.

6. Carolyn Johnsen. "The Wal-Marting of Agriculture," *Nebraska Nightly*, NPRN, 1 December 1998.

7. Johnsen, "Factory Hog Farms Defined."

8. Volland, Craig, "Analysis of Potential Impact of Furnas County Farms Hog Finishing Facility on the Municipal Wells of the Village of Orleans and the City of Alma, Nebraska," Spectrum Technologists, Kansas City, 6 October 1997.

9. Johnsen, "Local Governments Respond."

10. "Hog Site Starts Zoning Work by County Board," *Harlan County Journal*, 14 August 1997.

11. Carolyn Johnsen, "Hog Farm Trials Raise Constitutional Issues," *Nebraska Nightly*, NPRN, 29 June 2000.

12. Ann Brown, "Hog Factory Hits Snag in Harlan Co.," *Kearney Hub*, 8 October, 1997, 1A, 6A.

13. Volland, "Analysis of Potential Impact."

14. Johnsen, "Local Governments Respond."

15. Rick Calkins, testimony in *City of Alma v. Furnas County Farms et al.*, Harlan County District Court, case no. 5792, Bill of Exceptions, 27 June 2000, 186.

16. Ordinance nos. 10-077-1, 10-217-1, 10-217-3, 11-047-1, 11-047-2, and 11-047-3, City of Alma, October 1997.

17. Noyes W. Rogers, letter to Douglas R. Walker, 21 October 1997; Exhibit "H," *City of Alma v. Furnas County Farms et al.*

18. Noyes W. Rogers, letter to Douglas R. Walker, Exhibit "I," *City of Alma v. Furnas County Farms et al.*

19. Petition (5 November 1997) and Answer and Cross-Petition (18 November 1997), *City of Alma v. Furnas County Farms et al.*

20. Johnsen, "Local Governments Respond."

21. Johnsen, "Local Governments Respond."

22. Johnsen, "Local Governments Respond."

23. Jerry Schmitt, interview by author, Ord, Nebr., 28 March 2001.

24. Jerry Schmitt, Hearing Transcript, LB1152, "A Bill to Provide for County Interim Zoning Measures and Powers" (95th Nebr. Legis., 2d sess., Committee on Government, Military and Veterans Affairs), 4 February 1995, 3.

25. Don Stenberg and Timothy J. Texel, "Constitutional Considerations When Drafting Legislation Imposing a Temporary Moratorium on Construction of Hog Confinement Facilities," Attorney General Opinion 97044, 22 August 1997, *http://www.ago.state.ne.us*, retrieved 10 October 2002.

26. Ron Sedlacek, LB1152 Hearing Transcript, 4 February 1998, 89.

27. Carolyn Johnsen, "Hog Farm Opposition," *Nebraska Nightly*, NPRN, 4 December 1998.

28. Carolyn Johnsen, "Hog Farm Zoning Bill," *Nebraska Nightly*, NPRN, 4 February 1998.

29. *1998 Nebraska Unicameral Roster*, Patrick J. O'Donnell, Clerk of the Legislature.

30. Ernie Chambers, LB1152 Debate Transcript (95th Nebr. Legis., 2d sess.), 2 March 1998, 12386.

31. Don Preister, LB1152 Debate Transcript, 12379.

32. Bud Robinson, LB1152 Debate Transcript, 12361.

33. Roger Wehrbein, LB1152 Debate Transcript, 12335.

34. Curt Bromm, LB1152 Debate Transcript, 12405.

35. David Landis, LB1152 Debate Transcript, 13483.

36. Cap Dierks, LB1152 Debate Transcript, 13504.

37. Sally Herrin, "Act Now: Zoning is NOT Retroactive," *The Nebraska Report*, Nebraskans for Peace, March 1999, 6.

38. Johnsen, "The Wal-Marting of Agriculture."

39. Carolyn Johnsen, "Hog Water Hearing," *Nebraska Nightly*, NPRN, 1 June 1998.

40. John Opie, e-mail to author, 29 October 2001. Opie estimates 67 percent of the Ogallala lies beneath Nebraska. The "hidden treasure" characterization is from Opie's book *Ogallala: Water for a Dry Land* (Lincoln: University of Nebraska Press, 1993), 3.

41. Opie, *Ogallala*, 17.

42. Laura Jackson, "Water Quality," in *Understanding the Impacts of Large-Scale Swine Production: Proceedings from an Interdisciplinary Scientific Workshop*, ed. Kendall Thu, Des Moines, Iowa, 29–30 June 1995, 14. Sponsors: North Central

Regional Center for Rural Development, University of Iowa (UI) Environmental Health Sciences Research Center, Iowa's Center for Agricultural Safety and Health, The Farm Foundation, UI Center for Health Effects of Environmental Contamination.

43. Upper Republican NRD, "Groundwater Status," *http://www.urnrd.org*, retrieved 4 October 2002.

44. Johnsen, "Hog Water Hearing."

45. Settlement Agreements between Champion Valley Enterprises, L.L.C., and the Upper Republican Natural Resources District and between Steven and Tim Leibbrandt and the Upper Republican Natural Resources District, 7 July 1998.

46. *Hagan v. Upper Republican* NRD, Opinion of the Supreme Court of Nebraska, S-99-374 (2 March 2001).

47. Dawn Hansen, "New Zoning Will Restrict Feed Yards: Lincoln Co. Enacts Tough New Laws," *North Platte Telegraph*, 27 June 2000, A1.

48. Lindsay Young, "80 Plus Attend Polk Co. Hog Farm Hearing," *Columbus Telegram*, 7 March 2001, 1A.

49. Author's notes from the Gage County board's hearing on the Linsenmeyer permit, 30 May 2001.

50. Author's notes from the Holt County board's hearing on the Huston permit, 16 March 2001.

51. Carolyn Johnsen, "Rock County Rejects Hogs," *Nebraska News*, NPRN, 5 May 1999.

52. Author's notes from Nance County board hearing, 24 April 2001.

53. Dan Hodges, "National Report," *Nebraska Pork Talk*, April 1998, 8.

54. Keith Marvin, telephone interview by author, 7 May 2001.

55. Kendall Thu, telephone interview by author, 27 April 2001.

56. *Premium Farms v. County of Holt*, 263 Neb. 415 (2002).

57. *Furnas County Farms v. Hayes County et al.*, U.S. District Court for the District of Nebraska, 8:00CV548, Memorandum and Order, 14 September 2001.

58. *Coyote Flats, L.L.C. v. Sanborn County Commission*, 596 N.W. 2d 347 (S.D. 1999).

59. *Jeremiah Welsh v. Centerville Township*, 595 N.W. 2d 622 (S.D. 1999).

60. *Robert T. Richardson v. Township of Brady*, 2000 FED App. 0215 (6th Cir. 2000).

61. *Jeremy and Janice Borron v. Buck Farrenkopf et al.*, Missouri Court of Appeals, Western District (23 November 1999).

62. *Timothy H. Craig and the Chatham County Agribusiness Council v. County of Chatham, Chatham County Health Department, and the Chatham County Board of Health*, Supreme Court of North Carolina, no. 270PA01, 28 June 2002.

63. *Goodell v. Humboldt County*, 575 N.W.2d 486 (Iowa 1998).

5. THE LEGISLATURE WEIGHS IN

1. Randolph Wood, "State Is Protecting Water Quality," *Omaha World-Herald*, 22 October 1997, 15.

2. Randolph Wood, "A Report on Discussions of the Livestock Work Group and Agency Conclusions," NDEQ, Lincoln, Nebr., 12 January 1998, 9–10.

3. Wood, "Report of the Livestock Work Group," 11.

4. Randolph Wood, Hearing Transcript, LB1209, Livestock Waste Management Act (95th Nebr. Legis., 1st sess., Natural Resources Committee), 30 January 1998, 120.

5. NDEQ, Water Quality Division, *1996 Nebraska Water Quality Report* (Lincoln, April 1996), 77. Of Nebraska's 514 publicly owned lakes, 80 of the largest were assessed for the report. Of those, 74 were eutrophic or hypereutrophic.

6. Wood, "Report of the Livestock Work Group," 5–9.

7. Jon Bruning, interview by author, Lincoln, Nebr., 22 March 2001.

8. Ed Schrock, interview by author, Lincoln, Nebr., 26 February 2001.

9. Schmitt, interview.

10. Jim Jones, interview by author, Lincoln, Nebr., 26 March 2001.

11. Permitting file for Jones Enterprise Livestock Operation, Custer County, NDEQ Livestock Program.

12. Ardyce Bohlke, interview by author, Lincoln, Nebr., 21 February 2001.

13. Larry Williams, State Veterinarian, from interview in Johnsen, "Hog Carcass Disposal Questioned."

14. Don Preister, interview by author, Lincoln, Nebr., 9 March 2001.

15. Beutler, interview.

16. *Understanding the Impacts of Large-Scale Swine Production, Proceedings from an Interdisciplinary Scientific Workshop*, Des Moines, Iowa, 29–30 June 1995.

17. Carolyn Johnsen, "CLEAN Coalition," *Nebraska Nightly*, NPRN, 29 January 1998.

18. Jay Wolf, LB1209 Hearing Transcript, 30 January 1998, 36.

19. Randolph Wood, LB1209 Hearing Transcript, 30 January 1998, 115. Wood said "150 plus" permits were issued in 1997; Brian McManus of NDEQ confirms 158 permits issued that year.

20. Bryce Neidig, "Livestock Industry Under Siege," guest editorial, *Lincoln Journal Star*, 29 March 1998, 4D.

21. "Reports of Principals and Lobbyists," Clerk of the Legislature, 95th Nebr. Legis., 2d sess. (1998).

22. Bohlke, interview.

23. Roger Wehrbein, interview by author, Lincoln, Nebr., 22 March 2001.

24. Jerry Schmitt, LB1209 Debate Transcript, 30 March 1998, 14981.

25. NDEQ, "Lagoon Seepage Rates."

26. John C. Allen, Sam Cordes, Amy M. Smith, Matt Spilker, and Amber Hamil-

ton, "Environmental Issues and Perceptions of Rural Nebraskans," 1996 Nebraska Rural Poll, working paper, Center for Rural Community Revitalization and Development, University of Nebraska–Lincoln (UNL) Institute of Agriculture and Natural Resources, Research Report 96-3, August 1996.

27. Bob Ruggles, "Executive Director's Report," *Nebraska Pork Talk*, April 1998, 4.

28. NDEQ, "Plans Received '98," NDEQ database printout faxed to author by Brian McManus, 5 May 2001.

29. John Csukker, letter to Gary Buttermore, NDEQ, 24 June 1999; Environmental Services, Inc., *Manure/Nutrient Management Plan*, 1, in NDEQ permitting file for Enterprise Partners Arthur County Farrowing Site.

6. HOG HILTONS AND INITIATIVE 300

1. "Going by the Book, Hog Farm Taking Careful Steps to Meet Initiative 300 Rules," *Grand Island Independent*, 9 September 1999, 1.

2. Order, *Stenberg vs. Nebraska Premium Pork, L.L.C., and Premium Farms, L.L.C.*, District Court of Antelope County, Nebraska, case no. CI9941 (D26), 3 December 1999.

3. Settlement Agreements between Seaboard Corporation, Seaboard Farms, Inc., and SBD LLC and State of Nebraska, District Court of Antelope County, Nebraska, case CI0064 (D26) (August 2001).

4. Brian Mogenson, interview by author, Neligh, Nebr., 11 July 2001.

5. Mogenson, interview.

6. *Omaha National Bank v. Spire*, 389 N.W.2d 269 (Neb. 1986).

7. Howard Silber, "Partial Files Cloud Extent of Corporate Farms," *Omaha World-Herald*, 27 October 1982.

8. Larry Parrott, "Initiative 300 Approval Swells With Rural Vote," *Omaha World-Herald*, 3 November 1982.

9. Norris Alfred, "U.S.A., Inc.," *Polk Progress*, 11 February 1982, 3.

10. "8 States Use Statutes to Limit Land Buys," *Omaha World-Herald*, 12 September 1982, 12A.

11. Neal Oxton, interview by author, Lincoln, Nebr., 23 July 2001.

12. Chuck Hassebrook, interview by author, Lincoln, Nebr., 13 June 2001.

13. "Deadline Approaches in Family Farm Drive," *Lincoln Journal*, 18 June 1982.

14. Drey Samuelson, telephone interview by author, 24 July 2001.

15. Ann Toner, "Farm Vote Too Close to Call," *Lincoln Star*, 3 November 1982.

16. James Cunningham, "Sign the Farmers Union Petition," *The Catholic Voice*, 18 June 1982.

17. "Initiative 300 Should Draw 'No' Vote from Nebraska Electorate," *Lincoln Journal*, 8 September 1982, 10.

18. Samuelson, interview.

19. "You Did It!" *Nebraska Union Farmer*, July 1982, 1.

20. Associated Press, "Petition Opposition Coalition Feared," *McCook Gazette*, 9 September 1982.

21. Howard Silber, "Group Aims to Defeat Farm-Owner Proposal," *Omaha World-Herald*, 8 September 1982.

22. "Confusion and Silliness," editorial, *Omaha World-Herald*, 26 September 1982, 8M; "Think Again about Initiative 300," *Kearney Hub*, 6 October 1982; "Initiative 300 Should Draw 'No' Vote," *Lincoln Journal*.

23. "Proposition 300," editorial, *Burt County Plaindealer*, 30 September 1982, p. 6.

24. "2 of 3 Favored Initiative 300 in W-H Poll," *Omaha World-Herald*, 6 October 1982.

25. "A Summary of Political Campaign Financing," 1982 Primary and General Elections, Nebraska Accountability and Disclosure Commission, Lincoln, Nebr.

26. "Abstract of Votes," General Election, 2 November 1982, Nebraska Secretary of State, Lincoln, Nebr., 45.

27. Anne Toner, "Initiative 300 Passed by Voters," *Lincoln Star*, 3 November 1982, 1.

28. Ann Toner, "Legal Challenges to Farm Law Expected," *Lincoln Star*, 4 November 1982.

29. Samuelson, interview.

30. "Appeals to Emotionalism Worked for Initiative 300," editorial, *Omaha World-Herald*, 6 November 1982, 4.

31. Judge Donald E. Endacott, quoted in *Omaha National Bank v. Spire*.

32. MSM *Farms, Inc., v. Spire*, 927 F.2d 330 (8th Cir. 1991).

33. *Pig Pro Nonstock Co-op v. Moore*, 568 N.W.2d 217 (Neb. 1997).

34. *Norma L. Hall et al., v. Progress Pig, Inc.*, 575 N.W.2d 369 (Nebraska Supreme Court, 2000).

35. Allen Beermann, telephone interview by author, 26 September 2001.

36. "Settlement Agreement between the State of Nebraska and Christensen Family Farms, Inc.," Madison County District Court, case no. CI00-393, 23 March 2001.

37. William "Russ" Barger, interview by author, Lincoln, Nebr., 21 September 2001.

38. Beermann, interview; comment by Secretary of State Scott Moore in Carolyn Johnsen, "I-300 Enforcement," *Nebraska News*, NPRN, 22 June 2000.

39. Greg Lemon, telephone interview by author, 2 October 2001.

40. Betty Freese, "Pork Powerhouses 2000," *Successful Farming*, October 2000.

41. Rick Welsh, Chantal Line Carpentier, and Bryan Hubbell, "On the Effectiveness of State Anti-Corporate Farming Laws in the United States," *Food Policy*

26, no. 5 (October 2001): 543–48. This study was commissioned by Friends of the Constitution.

42. USDA NASS, *Agricultural Statistics Database.*

43. Jeffrey S. Royer and Roy Frederick, "I-300 Gets Thumbs Up from Majority of Nebraska Farmers in Recent Survey," *Nebraska Farmer*, August 1994, 12; John C. Allen, Rebecca Filkins, and Sam Cordes, "Rural Nebraska Tomorrow: The Gap between the Preferred and Expected Future," Center for Rural Community Revitalization and Development, UNL Institute of Agriculture and Natural Resources, Research Report 99-2, August 1999.

44. Alden Zuhlke, president of Nebraska Pork Producers Association, telephone interview by author, 2 October 2001.

45. Policy statements of the Nebraska Farm Bureau and the Nebraska Cattlemen, both adopted in December 2000, have nearly identical wording. The Farm Bureau statement was provided via e-mail by the Nebraska Farm Bureau's Governmental Relations Office. The Cattlemen's statement is found in "Resolutions and Policy Statements," *Nebraska Cattleman*, February 2001, 72.

46. George Beattie, interview by author, Lincoln, Nebr., 7 August 2001.

47. Norma Hall, interview by author, Elmwood, Nebr., 14 June 2001.

48. David Hendee, "In Hog Heaven," *Omaha World-Herald*, 7 April 1996.

49. Hassebrook, interview.

50. Marty Strange, telephone interview by author, 16 August 2001.

7. A TALE OF TWO COUNTIES

1. USDA NASS, *1997 Census of Agriculture.*

2. Lower Niobrara Natural Resources District, *Groundwater Quality Management Area*, January 1996, 2.

3. Charles Shapiro, William Kranz, Susan Olafsen Lackey, and Ralph Kulm, *Holt County Groundwater Education Project, Final Report* (O'Neill, Nebr.: Upper Elkhorn and Lower Niobrara NRDS, UNL, Natural Resources Conservation Service, and NDEQ, May 2001).

4. USDA NASS, *1997 Census of Agriculture.*

5. "National Farms Hog Operations Impact: $200 Million 'Ripple' in Area Economy," *Frontier and Holt County Independent*, 17 February 1983, 1.

6. "National Farms Hog Operations," 1.

7. Greg Gilsdorf, telephone interview by author, 23 June 2001.

8. William Haw, "Criteria for Selection of a New Feedlot," in *The Future of the Missouri Livestock Industry*, Special Report 349, Agricultural Experiment Station, University of Missouri–Columbia, 13–14 November 1986, 17.

9. "National Farms Head: Hog Facility Construction to Begin in Spring," *Frontier*

and *Holt County Independent*, 13 January 1983, 1–2; and "Concerns Over National Farms' Hog Operation Are Presented to Upper Elkhorn NRD Board Monday," *Frontier and Holt County Independent*, 20 January 1983.

10. Haw, "Criteria for Selection," 18.

11. Boyd Strope, letter to Lowell Creach, 30 November 1984, permitting file for National Farms (now Christensen Farms), NDEQ Livestock Program.

12. Dennis Heitmann, letter to Lowell Creach, 13 December 1984, National Farms permitting file, NDEQ.

13. Ken Lamb, letter to National Farms, Inc., 17 September 1985, National Farms permitting file, NDEQ.

14. *Goeke v. National Farms, Inc.*, 245 Neb. 262 (4 March 1994).

15. "Written Arguments Ordered, National Farms Hog Case," *Holt and Frontier County Independent*, 14 November 1991.

16. Gilsdorf, interview.

17. Wayne Kaup, interview by author, Stuart, Nebr., 21 June 2001.

18. John C. Allen and David J. Drozd, "Socioeconomic Impacts of Expanding Pork Production," University of Nebraska Cooperative Extension, RB336, March 2000.

19. William J. Weida, "A Brief Review of RB336: Socioeconomic Impacts of Expanding Pork Production," Department of Economics, Colorado College, Colorado Springs, 17 April 2000.

20. John Allen, e-mail to author, 19 July 2001.

21. Allen and Drozd, "Socioeconomic Impacts."

22. Jeff Gottula, memorandum to Gary Buttermore, 12 February 1999, permitting file for Christensen Farms, NDEQ Livestock Program.

23. Patrick W. Rice, letter to Robert A. Christensen, 17 January 2001, permitting file for Christensen Farms, NDEQ.

24. Mike Linder and Dennis Heitmann, interview by author, Lincoln, Nebr., 2 November 2001.

25. Haw, "Criteria for Selection."

26. USDA NASS, *Agricultural Statistics Database*.

27. Gary Olberding, interview by author, Stuart, Nebr., 20 June 2001.

28. Kelly Huston, interview by author, Emmet, Nebr., 16 March 2001.

29. Huston, interview.

30. USDA NASS, *Agricultural Statistics Database*.

31. Leon Riepe, interview by author, Beaver City, Nebr., 25 July 2001.

32. "Furnas County Potential Site for Large Swine Livestock Project," *Oxford Standard*, 30 November 1989.

33. "Consent Judgment Between Plaintiffs and Defendants Sand Livestock," 14 June 1989, and "Consent Judgment Between Cross-Plaintiffs and Cross-Defendants,"

19 June 1989, State of Michigan in the Circuit Court for the County of Ingham, case no. 86-57306-CE, Judge Thomas L. Brown.

34. Art Hovey, "Producer Has Outside Partners in Building of 2,000-Sow Units," *Lincoln Star*, 11 January 1990.

35. Sam Teply, manager of Lexington Livestock Market, telephone interview by author, 22 March 2002.

36. Gale Schafer of Sand Livestock Systems, Inc., letter to W. Clark Smith, NDEQ, 23 November 1992, permitting file for Gosper County Finisher in Frontier County, NDEQ Livestock Program.

37. Lillian Sayer, telephone interview by author, 22 March 2003; Clark Smith, memo to Randy Wood, Gale Hutton, and Brian McManus, 26 August 1994, permitting file for Furnas County Farms, NDEQ Livestock Program.

38. Riepe, interview.

39. Alan Thomas, LR123 Hearing Transcript, 16 September 1997, 130–31.

40. Steve Forbes's testimony to Chase County board, 28 August 2001, author's notes.

41. "Furnas County Hog Operations 'Not Perfect But Are Welcomed,'" *Keith County News*, 11 May 1997.

42. Mike Clements, Lower Republican NRD, e-mail to author, 24 October and 1 November 2001.

43. Megan Rabbass, "Hog Price Debate Takes Root in Nebraska," *Columbus Telegram*, 26 August 1997.

44. Furnas County Farms, letter to Citizens of Furnas County, Editor's Mail, *Oxford Standard*, 8 March 2001.

45. George Lauby, "Planners Oppose Mega-hog Request," *North Platte Telegraph*, 31 March 2001.

46. Michael J. Snyder, letter to Dwayne Fortkamp, 23 March 2001, Gosper County Road Dept., Elwood, Nebr.

47. Dick Reimer, letter to Clark Smith, 2 February 1993, permitting file for Gosper County Nursery, NDEQ Livestock Program.

48. Patrick Rice, warning letter to Dan Becker, 1 May 1997, permitting file for Gosper County Farrowing, NDEQ Livestock Program.

49. Tim Cumberland, letter to "Dear Nebraska Senator," 30 October 2001, Exhibit 6, Hearing Transcript, LR1285, "An Act to Create the Livestock Industry Task Force" (97th Nebr. Legis., 2d sess., Agriculture Committee), 12 February 2002.

50. Carolyn Johnsen, "Hog Farm Opponents Organize," *Nebraska Nightly*, NPRN, 4 December 1998.

51. Johnsen, "The Wal-Marting of Agriculture."

52. Carolyn Johnsen, "Small Farmers Try to Compete," *Nebraska Nightly*, NPRN, 9 December 1998.

53. Marty Strange, *Family Farming: A New Economic Vision* (Lincoln: University of Nebraska Press 1989), 171–72.

54. Allen and Drozd, "Socioeconomic Impacts."

55. John Ikerd, "Sustainable Agriculture, Rural Economic Development, and Large-Scale Swine Production," in *Pigs, Profits and Rural Communities*, ed. Kendall Thu and E. Paul Durrenberger (Albany: State University of New York Press, 1998).

56. Johnsen, "Hog Farm Opponents Organize."

57. Chuck Hassebrook, interview by author, in Carolyn Johnsen, "Schooley Obit," *Nebraska News*, NPRN, 17 May 2000.

8. THE MARSHAL COMES TO DODGE

1. Elaine Thoendel, interview by author, Ewing, Nebr., 4 June 2001.

2. Mogenson, interview.

3. Doug Rowse, telephone interview by author, 2 October 2001.

4. Donna Ziems, interview by author, Ewing, Nebr., 4 June 2001; Barger interview.

5. Ziems, interview.

6. Carolyn Johnsen, "Rock County Hogs," *Nebraska Nightly*, NPRN, 9 October 1998.

7. Loranda Daniels-Buoy and Lynda Buoy, interview by author, Bassett, Nebr., 9 February 2001.

8. "Game and Parks Concerned about Planned Hog Farms," *Grand Island Independent*, 5 January 1999.

9. Johnsen, "Rock County Hogs."

10. Deed of Trust with Future Advances and Construction Security Agreement, 26 February 1998, Antelope County Mortgage Records, Book 167, p. 263, Neligh, Nebr.

11. Betsy Freese, "Pork Powerhouses 1999," *Successful Farming*, October 1999.

12. Johnsen, "Rock County Rejects Hogs."

13. Johnsen, "Rock County Rejects Hogs."

14. Johnsen, "Rock County Rejects Hogs."

15. Jim Hanson, interview by author, Elsie, Nebr., 10 April 2001.

16. Charlynn Hamilton and Duane Fortkamp, letter to Michael Linder, 10 November 2000, Exhibit A in *Sand Livestock Systems, Inc., et al., v. Amy Svoboda et al.*, case no. C101-89, Platte County District Court, Petition and Praecipe, 20 February 2001.

17. Patrick W. Rice, memo to Commenters to Public Notice, Furnas County Farms–Enterprise North and South Finishing Site, 7 February 2001, permitting file for Enterprise Partners North and South Finishing Site, Hayes County, NDEQ Livestock Program.

18. *Sand Livestock Systems, Inc., et al. v. Amy Svoboda et al.*

19. Charlynn Hamilton, interview by author, Palisade, Nebr., 11 April 2001.

20. Tina Kitt, interview by author, Wauneta, Nebr., 27 June 2001.

21. Waterkeepers Alliance, "Hayes County Residents Fight Back Against Charges of Defamation by Sand Livestock Systems," Media Advisory, 8 October 2001.

22. Paul Hammel, "Kennedy Team Offers to Help With Lawsuit," *Omaha World-Herald*, 11 October 2001.

23. Nancy Thompson, telephone interview by author, 15 August 2001.

24. Tim Cumberland, letter to "Dear Nebraska Senator."

25. Karen Hudson, telephone interview by author, 20 August 2001.

26. Pat Rice, interview in Carolyn Johnsen, "Hog Lagoons Can Leak," *Nebraska Nightly*, NPRN, 3 December 1998.

27. Hanson, interview.

28. Carol Schooley, interview by author, Fullerton, Nebr., 13 March 2001.

29. NDEQ, "Nebraska to Receive $1.85 Million from IBP Agreement," news release, Lincoln, Nebr., 12 October 2001.

30. NDEQ, news release.

31. Richard Whiteing, interview by author, Jackson, Nebr., 18 June 2001.

32. Richard Whiteing, "Greed Leads to Big Hog Operations," *Columbus Telegram*, 2 October 1997, p. 4A.

33. Tim Cumberland, "Sand Responds to Hog Letter," *Columbus Telegram*, 5 October 1997, p. 4A.

34. Charles W. Sand Jr., Columbus, Nebr., letter to Archbishop Elden Curtiss, Omaha, Nebr., 10 June 1998.

35. Archbishop Elden Curtiss, letter to Mr. Charles W. Sand Jr., 24 June 1998.

36. Kendall Thu, interview.

37. Beutler, interview.

38. Schrock, interview.

39. John C. Allen, Rebecca Filkins, Sam Cordes, and Eric J. Jarecki, "Nebraska's Changing Agriculture: Perceptions about the Swine Industry," working paper, Center for Rural Community Revitalization and Development, UNL Institute of Agriculture and Natural Resources, Research Report 98-5, October 1998.

9. PORK TENDERLOIN AT THE CAPITOL

1. Carolyn Johnsen, "Pork Dinner," *Nebraska Nightly*, NPRN, 31 March 1998.

2. Johnsen, "Pork Dinner."

3. Johnsen, "Pork Dinner."

4. "Reports of Principals and Lobbyists," Clerk of the Legislature, 95th Nebr. Legis., 2d sess. (1998), and 96th Legis., 1st sess. (1999).

5. Carolyn Johnsen, "Waste Act Clarified," *Nebraska Nightly*, NPRN, 21 July 1998.

6. Brian McManus of NDEQ's Public Information Office provided the list of attendees.

7. Randolph Wood, memo to Governor Nelson, 3 November 1998, permitting file for Valley View Swine Producers of Cuming County, NDEQ Livestock Program.

8. Wood to Nelson, 3 November 1998.

9. Linder, interview.

10. Federal campaign figures are from the database of the Center for Responsive Politics, *http://www.opensecrets.org*, retrieved 21 and 22 September 2002.

11. Ben Nelson, "Report of Earmarked Contributions" (Financial Disclosure Statements), 1 April 1992 through 5 February 1999, on file at the Nebraska Accountability and Disclosure Commission, Lincoln, Nebr.

12. Nelson, "Report of Earmarked Contributions."

13. J. Christopher Hain, "Nelson's Past Motivates Crusade for Senate Seat," *Lincoln Journal Star*, 1 October 2000; Jake Thompson, "Ferocious Foe No Match for Senator," *Omaha World-Herald*, 26 August 2001.

14. "1998 Campaign Finance Summary," Nebraska Accountability and Disclosure Commission, Lincoln, Nebr.

15. Center for Responsive Politics, *http://www.opensecrets.org*, retrieved 28 October 2001 and 6 October 2002.

16. "1998 Campaign Finance Summary."

17. Art Hovey, "Livestock Waste Tops Candidates' Lists," *Lincoln Journal Star*, 9 July 1998, B1.

18. Mike Johanns, First Inaugural Address, 7 January 1999.

19. Mike Johanns, State of the State Address, 27 January 1999.

20. Buoy and Daniels-Buoy, interview; Jeri Kuchera, interview by author, Bassett, Nebr., 9 February 2001.

21. Cleve Trimble and Jim Jones, Campaign Statements, 2000 General Election, Nebraska Accountability and Disclosure Commission.

22. Jones, interview.

23. Mohammed Dahab, interview by author, Lincoln, Nebr., 31 May 2001.

24. Carolyn Johnsen, "Manure Task Force Reax," *Nebraska Nightly*, NPRN, 2 December 1998. Critics included Ron Schooley of Mid-Nebraska PRIDE and Nancy Thompson of the Center for Rural Affairs.

25. Jodi Thompson, telephone interview by author, 29 October 2001.

26. Cumberland, letter to "Dear Nebraska Senator."

27. Tina Kitt, "Let's Hope Our Reps in the State Lege Stand By Their Local Control Schtick" (editorial), *Wauneta (Nebr.) Breeze*, 17 January 2002, 2.

28. Russ Pankonin, "Commissioners Deny Special Use Permit for Hog Operation," *Wauneta Breeze*, 23 May 2002, 12.

29. George Lauby, "Furnas Farms' Pig Application Denied," *North Platte Telegraph*, 17 May 2002.

30. Susan Seacrest, telephone interview by author, 15 October 2001.

31. Charles Harness and Pat McConegle, memorandum to state executives, 2 January 1997.

32. Elizabeth Becker, "Unpopular Fee Makes Activists of Hog Farmers," *New York Times*, 11 June 2001, national edition, A1, A13.

33. Steve Cady, e-mail to author, 12 November 2001.

34. *Michigan Pork Producers et al. v. Campaign for Family Farms et al.*, United States District Court, Western Division of Michigan, Southern Division, Case No. 1:01-CV-34, Kalamazoo, Mich., 25 October 2002.

35. Bill Hord, "Pork Group Pushes Out Director," *Omaha World-Herald*, 8 January 2002, sunrise edition.

10. ANOTHER PASS AT THE LEGISLATURE

1. "Animal Confinement Policy National Task Force: Preliminary State Policy Survey Results," chaired by Mark A. Edelman, Iowa State University, Ames, 18 June 1999, *http://www.cherokee.agecon.clemson.edu/confine.htm*, retrieved 30 January 2001.

2. Schrock, interview.

3. Tom Osborne, Hearing Transcript, LB822, "Provide for Temporary Zoning Regulations and Prohibit Livestock Waste Control Facilities in Certain Watersheds" (96th Nebr. Legis., 1st sess., Natural Resources Committee), 25 February 1999, 4.

4. "Report on a Study Conducted by the LB1209 Livestock Waste Management Task Force and the Nebraska Department of Environmental Quality," NDEQ, Lincoln, Nebr., 1 December 1998, 13.

5. Gary Gausman, interview in Johnsen, "Hog Carcass Disposal Questioned."

6. Williams, interview, in Johnsen, "Hog Carcass Disposal Questioned."

7. Dr. Robert Wills, interview in Carolyn Johnsen, "Hog Carcasses/LB870," *Nebraska Nightly*, NPRN, 12, 13 April 1999.

8. Bruning, interview.

9. Bohlke, interview.

10. "Report on a Study Conducted by the LB1209 Livestock Waste Management Task Force and the Nebraska Department of Environmental Quality," 1 December 1998, 19.

11. Mike Linder, Hearing Transcript, LB870, "Change Provisions Relating to Livestock Waste and Environment Issues" (96th Nebr. Legis., 1st sess., Natural Resources Committee), 24 February 1999, 104.

12. Johnsen, "Hog Farm Stinks."

13. Johnsen, "Hog Farm Stinks."

14. Johnsen, "Hog Farm Stinks."

15. Johnsen, "Hog Farm Stinks."

16. *Earl P. and Kathleen M. Stephens et al., v. James D. Pillen et al.*, Answer and Counterclaim, case no. C100-69, District Court of Boone County, Nebr., 2 February 2001.

17. *Earl P. and Kathleen M. Stephens et al., v. James D. Pillen et al.*, Order, District Court of Boone County, Nebr., 31 October 2002.

11. BUILDING ON SAND

1. Jim Lawler, interview by author, Arthur, Nebr., 27 August 2001.

2. NDEQ Livestock Program permitting file for Enterprise Partners Arthur County Farrow Site, Form WP-42, 29 October 1998, 5.

3. Mary Pierce, "Expressing Their Concerns," *North Platte Telegraph*, 7 February 1999, A1.

4. Kent Miller, manager of Twin Platte NRD, telephone interview by author, 6 November 2001.

5. Lawler, interview.

6. Ron Lage, interview by author, Arthur, Nebr., 27 August 2001.

7. Mary Pierce, "Hogging the Spotlight: Opposition to Large-Scale Swine Farm Increasing," *North Platte Telegraph*, 15 January 1999.

8. U.S. Department of Commerce, Bureau of Economic Analysis, cited in "The Latest Data on Rural Poverty," Center for Rural Affairs Newsletter, July 2001, 1.

9. Dennis Heitmann, interview by author, Lincoln, Nebr., 2 November 2001.

10. Chuck Sand, fax to Governor Ben Nelson, 27 October 1998, NDEQ permitting file, Enterprise Partners Arthur County Farrow Site.

11. Gale Schafer, memo to Chuck Sand, 23 October 1998, NDEQ permitting file, Enterprise Partners Arthur County Farrow Site.

12. Mary Pierce, "Expressing Their Concerns," *North Platte Telegraph*, 7 February 1999, A1-2.

13. Larry Sitzman, letter to the *Lincoln Journal Star*, 28 February 1999.

14. Gary Buttermore, LWCF Memo to Pat Rice, 10 June 1999, NDEQ permitting file, Enterprise Partners Arthur County Farrow Site.

15. Jim Musilek, letter to Renee Hancock, 23 June 1999, and Gary Buttermore, memo to file, 24 June 1999, NDEQ permitting file, Enterprise Partners Arthur County Farrow Site.

16. Dennis Heitmann, letter to Gary Gausman, 25 June 1999, NDEQ permitting file, Enterprise Partners Arthur County Farrow Site.

17. Gary Buttermore, letter to Gary Gausman, 30 June 1999, NDEQ permitting file, Enterprise Partners Arthur County Farrow Site.

18. Gary Buttermore, "Summary of File Chronology, Enterprise Partners, Arthur County," 18 February 2000, NDEQ permitting file, Enterprise Partners Arthur County Farrow Site.

19. Gary Buttermore, memo to Dennis Heitmann, 8 October 1999, NDEQ permitting file, Enterprise Partners Arthur County Farrow Site.

20. Julie Anderson, "Wetlands Ordered Restored in Southwestern Nebraska," *Omaha World-Herald*, 1 September 1999.

21. "EPA Seeks $70,000 Penalty for Wetlands Damage in Perkins County, Nebraska," EPA Region 7 Press Release, 18 October 2000; Becky Uehling, "EPA Files Motion to Withdraw Complaint Against Hog Facility, *Grant Tribune-Sentinel*, 5 April 2001, 3.

22. Ron Lage, letter to Michael Linder, 8 March 2000, NDEQ permitting file, Enterprise Partners Arthur County Farrow Site.

23. Dennis Heitman letter to Gary Gausman, 1 November 1999, NDEQ permitting file, Enterprise Partners Arthur County Farrow Site.

24. Ron Lage, interview in Carolyn Johnsen, "Lagoons Eroded," *Nebraska News*, NPRN, 23 March 2000.

25. Linder, interview.

26. Linder, interview.

27. Tom Arrandale, "A Tale of Two EPAS," *Governing*, June 1998, 40.

28. Deb Gilg, letter to Patrick W. Rice, 23 March 2000, NDEQ permitting file, Enterprise Partners Arthur County Farrow Site.

29. Pat Rice, "Notice of Violation" letters to Gary Gausman, 23 March 2000, NDEQ permitting file, Enterprise Partners Arthur County Farrow and Perkins County Nursery.

30. Michelle Stirek, letter to Natural Resources Committee, Nebraska Legislature, 24 March 2000, NDEQ permitting file, Enterprise Partners Arthur County Farrow.

31. Rate calculated from Enterprise Partners Form WP-42 and NDEQ, "Lagoon Seepage Rates."

32. "NRCS Review of the Enterprise Partners Ag Waste Facility Arthur County," NDEQ permitting file, Enterprise Partners Arthur County Farrow Site.

33. Kent Miller, telephone interview by author, 6 November 2001.

34. Nitrate levels in Lawler's well taken from "Water Analysis Report," Olsen's Agricultural Laboratory, Inc., McCook, Nebr., 22 January 1999, 14 April 2000, and 5 June 2001.

35. Michael Jess, interview by author, Lincoln, Nebr., 2 July 2001.

36. Miller, interview.

37. Randolph Wood, LB1209 Hearing Transcript, 30 January 1998, 119.

38. Joe Duggan, "Does Dairy Threaten Stream?" *Lincoln Journal Star*, 23 January 2000, 1A–5A.

39. Marty Link, interview by author, Lincoln, Nebr., 5 November 2001.

40. Sadayappan Mariappan, *Impact of Lagoon Leakage at Confined Animal Feeding Operations in Nebraska on Shallow Ground Water Nitrate Concentrations and N-Isotope Variability* (unpublished master's thesis, University of Nebraska–Lincoln, December 2001).

41. Roy F. Spalding, interview by author, Lincoln, Nebr., 8 August 2001.

42. Link, interview.

43. Enzo R. Campagnolo and Carol S. Rubin, "Report to the State of Iowa Department of Public Health on the Investigation of the Chemical and Microbial Constituents of Ground and Surface Water Proximal to Large-Scale Swine Operations," Final Draft (Atlanta: National Center for Environmental Health, Centers for Disease Control and Prevention, October-December 1998).

44. Campagnolo and Rubin, "Report of Chemical and Microbial Constituents."

45. Stuart Leavenworth, "Tests Link Tainted Wells to Farms," *Raleigh News & Observer*, 21 August 1998, 1A.

46. J. M. Ham, L. N. Reddi, C. W. Rice, and J. P. Murphy, "Evaluation of Lagoons for Containment of Animal Waste," Department of Agronomy, Kansas State University, Manhattan, Kans., 28 April 1998.

47. Roger Myers, "Study: Hog Lagoons Pass Seepage Test," *Topeka Capital-Journal*, 30 April 1998, C1.

48. Rick Koelsch, interview by author, Lincoln, Nebr., 11 June and 17 August 2001.

49. "Groundwater Sampling Results," *The Oracle* (O'Neill, Nebr.: Upper Elkhorn NRD), summer/fall 2000, 4.

50. Dennis Schueth, interview by author, O'Neill, Nebr., 19 and 22 June 2001.

51. NDEQ, Water Quality Division, *2000 Nebraska Water Quality Report* (Lincoln, 2000), 1–2, 146.

52. NDEQ, *2000 Nebraska Water Quality Report*, 3.

53. Jackson, "Water Quality," 37.

54. EPA, "1998 National Water Quality Inventory," cited in "Proposed Regulations to Address Water Pollution from Concentrated Animal Feeding Operations," EPA 833-F-00-016, March 2001.

55. Agriculture Director Merlyn Carlson, letter to EPA Office of Water, Engineering and Analysis Division, 27 July 2001, Lincoln, Nebr.

56. Governor Benjamin Nelson, news release, "Nelson Says Federal Livestock Waste Mandates Unnecessary," 12 June 1998, Lincoln, Nebr.

57. NDEQ Deputy Director Jay Ringenberg, letter to EPA Office of Water, Engineering and Analysis Division, 30 July 2001, Lincoln, Nebr.

58. "Comprehensive Study of Water Quality Monitoring in Nebraska," LB1234 Report, Phase I, 1 December 2000, and Phase II, August 2001 (LB1234, "Create the Niobrara Council and the Ethanol Pricing Task Force; Provide for Wastewater Treat-

ment Facility Grants and a Water Quality Monitoring Study; and Prohibit the Sale of MTBE," 96th Nebr. Legis., 2d sess., Natural Resources Committee).

12. THE SMELL OF MONEY

1. Janie Mullanix, interview by author, Champion, Nebr., 12 April 2001.

2. Carolyn Johnsen, "Local Governments Respond."

3. Loral Johnson, "Chase County Hosts Urban/Rural Ag Understanding Tour," *Imperial Republican*, 22 July 1999.

4. Shona Heim, interview by author, Champion, Nebr., 12 April 2001.

5. Mullanix, interview.

6. LB1209 Livestock Waste Management Task Force Study, 19.

7. Stan Schellpeper, LB1209 Debate Transcript, 30 March 1998, 14980.

8. Dave Maurstad, LB1209 Debate Transcript, 30 March 1998, 15001.

9. Donham, "Respiratory Disease Hazards," 50.

10. Donald Leopold, interview by author, Omaha, Nebr., 22 August 2001.

11. Stewart Melvin, "Air Quality," in *Understanding the Impacts of Large-Scale Swine Operations*, 47.

12. Donham, "Respiratory Disease Hazards," 49–58.

13. Koelsch, interview.

14. Donham, "Respiratory Disease Hazards," 50.

15. Nebraska Administrative Code, "Ambient Air Quality Standards," Title 129, Chapter 4, .007 "Total Reduced Sulfur."

16. Chris Ison, "State Health Department Acknowledges Health Risks of Feedlots," *Minneapolis Star Tribune*, 20 February 2000.

17. Ison, "State Health Department Acknowledges Health Risks of Feedlots."

18. Wayne Gieselman of Iowa DNR, telephone interview by author, 27 February 2002.

19. Iowa State University and the University of Iowa Study Group, *Iowa Concentrated Animal Feeding Operations Air Quality Study: Final Report* (Des Moines: Iowa Department of Natural Resources, February 2002).

20. Odor Control Task Force, "Control of Odor Emissions from Animal Operations: A Report from the Board of Governors of the University of North Carolina," (North Carolina Agricultural Research Service, College of Agriculture and Life Sciences, North Carolina State University), 1 September 1998.

21. Paul Hammel, "Study Ties Emissions, Illnesses in South Sioux City," *Omaha World-Herald*, 9 May 2001.

22. Steven Inserra, "Evaluation of Neurobehavioral Health Status for Chronic and Repeated Exposure to Hydrogen Sulfide, Dakota and South Sioux Cities, Nebraska"

(Atlanta: Agency for Toxic Substances and Disease Registry, U.S. Department of Health and Human Services, September 2002).

23. Thu, Donham, Ziengenhorn, Reynolds, Thorne, Subramanian, Whitten, and Stookesberry, "Control Study of Residents Living Near a Large-Scale Swine Operation."

24. Steve Wing and Susanne Wolf, "Intensive Livestock Operations, Health and Quality of Life among Eastern North Carolina Residents," *Environmental Health Perspectives* 108, no. 3 (March 2000): 233–38.

25. Susan S. Schiffman, Elizabeth A. Sattely-Miller, Mark S. Suggs, and Brevick G. Graham, "Mood Changes Experienced by Persons Living Near Commercial Swine Operations," in *Pigs, Profits and Rural Communities*, 84–102.

26. Susan S. Schiffman, John M. Walker, Pam Dalton, Tyler S. Lorig, James H. Raymer, Dennis Shusterman, and C. Mike Williams, "Potential Health Effects of Odor from Animal Operations, Wastewater Treatment, and Recycling of Byproducts," *Journal of Agromedicine* 7, no. 1 (2000): 16.

27. Susanna Von Essen, telephone interview by author, 31 October 2001.

28. George Lauby, "Group Considers Downside of Factory Hog Farms," *North Platte Telegraph*, 1 August 2001, A3.

29. Charley Colton, interview by author, Imperial, Nebr., 27 June 2001.

30. Mogenson, interview.

31. Tom and Elaine Kimes, interview by author, Clearwater, Nebr., 22 June 2001.

32. Linder, interview.

13. TO MAKE A SILK PURSE OUT OF A SOW'S EAR

1. Kaup, interview.

2. USDA NASS, *Agricultural Statistics Database*.

3. NDEQ, Water Quality Division, *2002 Nebraska Water Quality Report* (Lincoln, September 2002).

4. Paul Hammel, "Turning Hog Odors into Tax Deductions," *Omaha World-Herald*, 19 May 2002, 1.

5. Paul Hammel, "Hog-Farm Owner's Suit Alleges Cousin Slandered Him, Lot Plan," *Omaha World-Herald*, 5 March 2002, 8B.

6. *South Dakota Farm Bureau, Inc., et al. v. Joyce Hazeltine et al.*, United States District Court, District of South Dakota, Central Division, CTV 99-3018 (16 May 2002); and South Dakota Attorney General Mark Barnett, "Amendment E Found Unconstitutional," news release, 17 May 2002.

7. Roy Frederick, "Livestock Producers Could Benefit from Environmental Program," *http://www.NebraskaStatePaper.com*, retrieved 25 May 2002.

8. John Austin, "Population Decline Characterizes Nebraska's Small Counties," UNL Bureau of Business Research, *Business in Nebraska* 57, no. 665 (March 2002).

9. Allen, Filkins, and Cordes, "Rural Nebraska Tomorrow."

10. Daniel B. Wood, "Coming Soon to a City Near You: A Farm," *Christian Science Monitor*, 3 January 2001, 1, 4.

11. Margaret Mellon, Charles Benbrook, and Karen Lutz Benbrook, *Hogging It: Estimates of Antimicrobial Abuse in Livestock*, (Cambridge, Mass.: Union of Concerned Scientists, 2001), xiii.

12. Farmers Market information, *http://www.ams.usda.gov/farmersmarkets/facts. htm*, retrieved 19 November 2001.

13. USDA NASS, *1997 Census of Agriculture*.

14. USDA NASS, *1997 Census of Agriculture*.

15. Paul Willis, interview by author, Thornton, Iowa, 7 September 2001.

16. Bill Niman, interview by author, Des Moines, Iowa, 8 September 2001.

17. Marsha McBride, interview by author, Thornton, Iowa, 7 September 2001.

18. Willis, interview.

19. Cal Peternell, interview by author, Thornton, Iowa, 7 September 2001.

20. Nicholas Lemann, "Toque Envy," *The New Yorker*, 24 September 2001, 90.

21. Bob Spenner, interview by author, West Point, Nebr., 12 November 2001.

22. Niman, interview.

23. Max Waldo, interview by author, DeWitt, Nebr., 9 August 2001.

24. Stan Rosendahl, telephone interview by author, 5 November 2001.

25. Worldwatch Institute, "United States Leads World Meat Stampede," press briefing on the global trends in meat consumption, 2 July 1998.

26. Delgado, Rosegrant, and Meijer, "Livestock to 2020."

Index

Adams County, 13, 65

Agency for Toxic Substances and Disease Registry (ATSDR), 132

Agriculture Committee (Nebraska legislature), 24, 45, 97

air quality and confined hog farms: complaints about, 39, 70–71, 76; control/lack of control of, 109–11; effects of, 2–3, 126–34; methods for management of, 48, 133–34; and private records of odors and symptoms, 125; and "smell of money," 19, 110; state regulation of, 3–4, 101; and studies of smells and odors, 19–20, 26

Albion NE, 19, 25

Alfred, Norris, 58

Allen, John, 72

Alma NE, 14, 30–31, 32, 42, 138

Ambrosek, Robert, 38

American Agriculture Movement, 60

American Medical Association, 141

anaerobic lagoons, 6–7, 53, 117–23, 130

Animal Health Institute, 141–42

Animal Welfare Institute, 145, 146

Antelope County, 14, 26, 27, 31, 83, 84, 87, 134

antibiotics, 5–6, 20, 141–42

anti-hog farm activism, 84–94, 99, 139. *See also under specific organizations*

Arapahoe NE, 77

Area Citizens for Resources and Environmental Concerns (ACRES), 88, 89, 94, 138

Arp, Susan, 25

Arthur County, 54, 112, 114, 115, 117–18

artificial insemination, 5

Atkinson NE, 13, 69

Audubon Nebraska, 106

Audubon Society, 35

Baker, Larry, 110

Barger, William "Russ," 64

Bassett NE, 33, 40

Bauerle, David, 126, 133

Bauerle, Dirk, 126

Bauerle, Mike, 126, 127

Beaver City NE, 77

Beaver Valley Reproductive Center, 4
Beermann, Allen, 59, 64
Bell, Rich, 21, 22, 50
Bell Family Farms: lobbying by, 51, 96, 97;
    sites for confinement operations of, 14, 19,
    20, 22, 37, 40, 54, 126–27
Bernard, George, 2
Bernard, Joyce, 2, 3
Bernard, Mabel, 1–4, *80-3*, 91, 130, 137
Beutler, Chris, 23, 48, 49, 50, 52, 93, 96, 97, 99,
    106, 109
biosecurity in hog production, 5, 9
Bloomfield NE, 24
Blue River, 137
Bohlke, Ardyce, 24, 47, 48, 49–50, 51, 52, 53,
    109
Boone County, 23, 34, 110, 111, 136
"Boss Hog," 14–15
Bowman, Mike, 50
Bowman Family Farms, 37, 54
Boyd County, 91
Brazil, 147
Broken Bow NE, 100
Bromm, Curt, 36
Brown County, 85, 100
Bruning, Jon, 46, 108–9
Brunswick NE, 14, 26
Bucktail Lake, 114
Buoy, Lynda, 86, 139
Burkey, Syd, 11

C22, 5
Cady, Steve, 104
Calkins, Rick, 32
Callaghan, Ken, 145
Carlson, Merlyn, 124
Cather, Willa, 91
Catholic Voice, 60
Cedar County, 99, 137
Cedar Rapids Finisher, 4
Cedar River, 18, 85
Center for Rural Affairs: advocating small
    family farms by, 52, 81, 82; criticism of cor-
    porate hog farms by, 16, 49, 89, 103; pro-
    moting I-300 by, 59, 60, 67
center-pivot waste disposal, 39, 44, 107, 123
Centers for Disease Control (CDC), 121, 141

Central City NE, 19
Central Farmers Co-op, 84
cesspits, 3
Chalmers, Rob, 145
Chambers, Ernie, 25, 36
Champion NE, 133
Champion Valley Enterprises, 125, 126, 133
Champion Valley NE, 101
Chase County: quality of air and water in,
    101, 125, 126–27, 133; sites of confinement
    hog production in, 1, 4, 22, 54, 77, 102; zon-
    ing process in, 37
Chase County Board of Commissioners, 78,
    101, 136
Chase County Planning Commission, 78
checkoff, 102–4
China, 147
Christensen Family Farms, 64, 65, 72
Citizens Against Virtually Everything (CAVE),
    93
Citizens for Air, Resources, and Environment
    (CARE), 86, 87, 139
Citizens for Environmental Stewardship, 91
Class A streams, 106
Clay County, 23
Clean Air Act, 91
Clean Water Act, 23–24, 46, 124
Clearwater Creek, 85
Coalition for Livestock, the Environment,
    and Agriculture in Nebraska (CLEAN), 49
Colorado, 37, 53, 66, 89
Colton, Charley, 133
Columbus NE, 4, 8, 19, 65, 75
Columbus Telegram, 92
Committee to Preserve the Family Farm, 59,
    62
compliance assistance, 117
Concerned Citizens for the Cedar Valley, 136
confined animal feeding operations (CAFOS),
    105
consumer selection of pork, 144–46
Corn Growers, 27
Crete NE, 8, 24, 26
cryptosporidium, 121
Cumberland, Tim, 26, 78, 88, 89, 92, 101, 102
Cuming County, 36, 132–33
Curtiss, Eldon, 92

Dahab, Mohammed, 101
Dakota City NE, 91, 132
Daniels-Bouy, Loranda, 86, 139
Dawson County, 63
dead pigs/carcasses, disposal of, 20, 26, 47, 53, 70, 107–8
dead pits, 26
"Death Valley," 30
DeCoster, A. J., 87
DeWitt NE, 95
Diamond B Beef, 146
Dickey, Robert, 99
Dierks, Cap, 37, 52, 97, 99
disease, in confined hog operations, 20
Dixon County, 23
Douglas, Paul, 58
Drabenstott, Mark, 12
Dubas, Annette, 10–11, 20, 21, 23, 26, 82, 90, 139
Dubas, Emil, 19, 20, 21, 22, 137
Dubas, Ron, 10–11, 13, 20, 21
Dubuque IA, 85
Dundy County, 2, 3, 37, 54
Durocs, 5, 74, 95

E. coli, 121
economies of scale, 16
Eddyville NE, 47
Elgin NE, 84
Elkhorn River, 84, 85, 137
Elm Creek NE, 106
Elmwood NE, 67
Enslen, Richard, 104
enterococcus, 121
Enterprise Partners: applications for permits by, 42, 54, 87–88, 133; environmental compliance of, 114–19; sites for confined hog operations of, 37, 113
Environmental Quality Council, 120
Environmental Quality Incentive Program (EQIP), 138
EPA. See U.S. Environmental Protection Agency
"Erin Brockovich" syndrome, 93
erysipelas, 20
European Union (EU), 141

Families Against Rural Messes (FARM), 89
family farming: and definition of "family," 59; numbers of hog farmers in, 8, 13, 136; trends and challenges to, 11, 21, 58–67
Family Farm Preservation Act, 59–67
Family Quality Pork Processors, 146, 147
Farmers Choice, 146
Farmers' Hybrid, 144
Farmland, 8, 57, 76, 147
Farm Lobby Day, 99
farrowing, 6, 9
Federal Election Commission, 98
Federal Reserve Bank of Kansas City, 12
feeder pigs, 6, 10
finishing pigs, 6, 80-6
First National Bank (Fullerton NE), 19, 21
Forbes, Steve, 77
4-H'ers, 74, 140–41
Frederick, Roy, 16–17
Frenchman Creek, 78
Frenchman River, 126
Friends of the Constitution, 64, 67, 100, 139
Friese, Kurt, 145
Frontier County, 76
Fullerton NE, 9, 18–22, 24, 92
Furnas County, 23, 68, 74–78, 140–41
Furnas County Farms: applications for permits by, 30–33, 41, 77, 78, 101, 136; compliance with environmental law by, 88, 102; litigation with Holt County by, 42; waste spills from, 76

Gage County, 40
Gausman, Gary, 33, 107–8
genetic manipulation, 5
Gilsdorf, Greg, 71
Glickman, Dan, 103
Gosper County, 76, 78
Gottsch, Bob, 99
Gottsch, Brett, 99
Governing Magazine, 117
GRACE Family Farm Project, 133
Grand Island NE, 62
Grange, 60
Great Depression, 29, 55
Great Plains, 12, 38
Greeley County, 14, 22, 23, 78

groundwater. *See* water resources
Groundwater Foundation, 102
Gulf of Mexico, 124
Guymon OK, 84

Haecker, Hal, 31
Hall, Norma, 67
Hall v. Progress Pig, 63
Ham, Jay, 122
Hamilton, Char, 88
Hampshires, 5
Hannon, Edward, 71
Hanson, Jim, 88, 90
Hardenburger, Phil, 97
hardship variance, 97
Harlan County, 14, 23, 29–33, 42
Harlan County Lake, 30, 32
Hassebrook, Chuck, 59, 67, 82
Hastings Correctional Center, 27
Hastings Pork, 13, 65
Haw, Bill, 70, 72
Hayes County, 42, 78, 88, 135
Heim, Shona, 126, 127, 133
Heineman, Dave, 98
Herrin, Sally, 37
Hidden Valley Pork, 25
Hines, Gary, 145
Hodges, Dan, 16, 41
hog cholera, 20
Hog Hilton, 57
hog prices, 6, 54–55, 105
"hog wars," 90
Holt County: applications for permits in, 22,
    31, 41, 83, 84, 87; corporate hog farms in,
    13, 40, 68–74, 80-6, 90; management of hog
    waste in, 122, 135–36, 138
Homestead Act, xi
Howard County, 22
Hudson, Karen, 89–90
Humboldt County, 43
Huston, Kelly, 40, 73–74
Huston, Marge, 73

IBP, 8, 76, 91, 147
Ikerd, John, 17
Imperial NE, 1, 37, 125, 126–27
Imperial Republican, 37

Initiative 300 (I-300), 16, 57–67, 69, 85, 91, 138
interim zoning, 34–43
Iowa, 48, 53, 58, 132, 142–44
Iowa Citizens for Community Improvement,
    103, 131
Iowa Department of Natural Resources, 43,
    131, 144
Iowa Department of Public Health, 121
Iowa Select, 142
Iowa State University, 131
Iowa Supreme Court, 43

Jansen, Julie, 131
Jarrett, Sue, 89
Jefferson, Thomas, 21
Jess, Michael, 119
Johanns, Mike, 98, 99
Johnson, Steve, 145
Jones, Jim, 47, 99–100

Kansas, 43, 58
Kansas State University, 122
Kaup, Kurt, 135
Kaup, Wayne, 80-3, 135–36
Kearney NE, 60
Kennedy, Robert, Jr., 89
Kerrey, Bob, 23, 60
Kimes, Elaine, 134, 137
Kimes, Tom, 134, 137
Kitt, Tina, 88, 102
Klassen, Lawrence, 18
Knopik, Carolyn, 20, 21, 22, 23
Knopik, Jim, 20, 21, 22, 23, 139
Knox County, 23
Koelsch, Rick, 12, 122
Kuchera, Geri, 87

Lage, Ron, 80-1, 114, 116–17
lagoons, 6–7, 53, 117–23, 130
Lake, James, 59
Lamb, Charles, 147
Landis, David, 36
Landrace breed, 5
Lawler, Jim, 80-1, 112–14, 118–19
Lawler, Peggy, 113
LB822, 106–7
LB1152, 34, 35, 36–37, 51

LB1193, 64
LB1209, 50–54, 100, 105, 107–9, 127
LeDioyt, Glenn, 61
Leibbrandt, Steve, 2, 4, 126, 133
Leibbrandt, Tim, 2, 4, 126, 133
Leopold, Donald, 128–30
Lesiak, Tony, 19, 21, 22
Lexington Livestock Market, 76
Lincoln County, 39, 120, 138
Lincoln Journal, 61
Lincoln Journal Star, 51, 80-9, 115
Lincoln NE, 84
Linder, Mike, 97, 116, 117, 134
Lindgren, Jeff, 26
Lindsay NE, 19
Linsenmeyer, Rod, 40
Livestock Waste Management Act, 50, 95, 99, 105, 115
Long Pine Creek, 33, 85, 86
Long Pine NE, 33, 106
Loup River, 18, 85, 137
Lower Republican NRD, 77

Maddux, Jack, 62
Madison NE, 8
Matzke, Gerald, 117
Maurstad, Dave, 127
McBride, Marsha, 144–45
Metropolitan Life, 61
Michigan, 42, 75, 88
Mid-America Dairymen, 97
Middle Logan Creek, 137
Mid-Nebraska PRIDE (People Responding in Defense of Our Environment): and concerns with confinement hog operations, 11, 27, 28, 35, 46, 49, 79, 82, 85, 90, 92, 94, 139; lobbying by, 51, 52; organization of, 21, 22, 23
Miller, Kent, 118, 119
Minnesota, 53, 58, 131
Minnesota Department of Health, 131
Mississippi River, 124
Missouri, 42, 58, 66
Missouri Rural Crisis Center, 103
Modlin, Linda, 91
Mogenson, Brian, 14, 50, 56–57, 83–87, 80-4, 134

Mogenson, Harry, 56, 87
Moore, Russell, 145
Moore, Scott, 98
Morris, Wright, 91
Morrow, Jeremy, 145
"mortgage busters," 11
Mount Echo, 4
MSM Farms, Inc., 63
Mueller, Bill, 97
Mullanix, Janie, 126–28, 137
Mullanix, Nicholas, 80-9, 130
Murphy, Wendell, 14
Murphy Family Farms, 21, 57

Nabor, Roland, 96
Nance County: citizen concerns with corporate hog operations in, 22, 23, 28, 31; contaminated water in, 137; sites for corporate hog operations in, 14, 19, 20; small hog productions in, 9
Nance County Planning and Zoning Commission, 40–41, 90
National Bank Act, 62
National Catholic Rural Life Conference, 28
National Farmers Organization, 60, 62
National Farmers Union, 103
National Farms, 13, 14, 40, 65, 69–74, 80-6, 138
National Pork Producers Council (NPPC), 5, 20, 103, 134
Natural Resources Committee (Nebraska legislature), 24, 45; on hog waste on cropground, 48; on NDEQ permit process, 93, 97, 108, 118; work on LB1209, 50
Natural Resources Conservation Service, 118
Natural Resources Defense Council, 23
NDEQ. *See* Nebraska Department of Environmental Quality
Nebraska Bankers Association, 49, 67, 97, 100
Nebraska Cattlemen, 27, 34, 49, 50, 51, 52, 66, 97, 100, 120
Nebraska Chamber of Commerce and Industry, 27, 34–35, 49, 100
Nebraska Constitution, Article 12, Section 8, 62
Nebraska Department of Agriculture, 16, 108
Nebraska Department of Environmental Control (NDEC), 70

Nebraska Department of Environmental
Quality (NDEQ): air quality records by, 91,
131; applications for permits processed by,
14, 22–25, 30–33, 53, 88, 96–99, 109; assessment of agricultural runoff by, 123, 124;
enforcement of state laws by, 27, 48, 49, 83,
90, 107, 114–21, 137–38; groundwater monitoring by, 45, 72, 77, 78, 120–21; odor management requirements of, 134
Nebraska Department of Water Resources,
119
Nebraska Farm Bureau: on confined hog
operations, 25, 27; lobbying by, 34, 35, 46;
opinions of, on regulation of corporate
hog farms, 49, 51, 52, 66, 97, 120; political
contributions of, 100
Nebraska Farmers Union: lobbying by, 35;
protecting family farms by, 52, 60, 61, 62;
on regulations of corporate hog farms, 25,
27, 37, 49, 58–59
Nebraska Pork Producers Association (NPPA),
99, 103–4; on consumers selecting pork
products, 147; lobbying by, 34, 95–96; on
regulating confinement hog operations, 23,
41, 49, 51, 52, 54, 66
Nebraska Premium Pork, 56
Nebraska Rural Poll, 53–54, 94, 139
Nebraska Sandhills, 54, 58, 62, 67, 85, 99, 112–
24
Nebraska Supreme Court, 13, 33, 42, 57, 62,
63, 67, 71, 137, 138
Nebraska Worth Fighting For, 84, 94
Neidig, Bryce, 51
Neligh NE, 14
Nelson, Ben, 27, 45, 46, 64, 96–99, 115, 124,
139
New England Journal of Medicine, 141
New River NC, 15
Niman, Bill, 142, 144–47
Niman Ranch Pork, 142, 146
Niobrara River, 85
nonfamily corporations, trends of, 58–67
"No on Initiative 300" Committee, 61–62
North Carolina: hog farm waste management
in, 12, 14, 48, 121–22; regulating hog opera-
tions in, 15, 42, 131, 138; study of hog farm
effects on humans, 132

North Carolina Department of Agriculture,
15
North Dakota, 37, 58
North Platte NE, 113, 114
North Star Neighbors, 11, 146
Nowka, Trent, 97, 98
nutrients and algal blooms, 45–46, 123

Oceanview Farm, 15
Ogallala NE, 37
Ogallala Aquifer, 38, 54, 68, 112
Ohrt, Ken, 30
Oklahoma, 58, 84
Olberding, Gary, 73
Omaha Chamber of Commerce, 49
Omaha National Bank, 62
Omaha World-Herald, 45, 61, 62
O'Neill NE, 8, 68, 69
Opie, John, 38
Orleans NE, 30–31, 32, 80
Ortmeier, Stan, 97
Osborne, Tom, 98, 106, 139
Owens, Michael, 111
Oxton, Neal, 58–59

Pape, Pam, 31
Parma MI, 75
Pawnee Pride Meats, 146
P.C. West Reproductive Center, 4, 7
Perkins County, 37, 42, 88, 116, 117–18
Peternell, Michael "Cal," 145
Petersburg NE, 146
Pheasant Ridge, 4
Pierce County, 23
pigs, bias against, 114
Pillen, Jim, 4–8, 23, 25, 26, 50, 80, 98–99, 110,
111, 136
Platte County, 98, 136, 137
Plattsmouth NE, 52
Podraza, Lory, 145
Polk County, 40, 138
Polk NE, 137
Polk Progress, 58
pollution. See air quality and confined hog
farms; water resources
"pollution shopping," 43
"poop police," 89–90

Pork Producers, 27

poultry growers, 34

Preister, Don, 36, 47–48, 99

Premium Farms: activism against, 26–27, 139; applications for permits by, 40–42, 56–57, 137; investigations of, 64; lobbying by, 51, 52; private citizen monitoring of, 83–87; sites for confined hog operations of, 14, 33, 37

Premium Standard Farms, 66

PRIDE. See Mid-Nebraska PRIDE

Progressive Swine Technologies, 5, 22, 34, 51, 80, 80-8, 96, 136

Progress Pig, 67

Prudential Insurance Company, 58, 61–62

pseudorabies, 20

Raikes, Ron, 106

Raleigh (NC) News and Observer, 14, 15

Red Willow County, 78, 136

Republican River, 29–30

respiratory diseases, 20

Riepe, Leon, 74–75, 76

Rinehart, Barb, 80-1

Robinson, Bud, 36

Rock County, 40, 85, 86, 87, 100, 139

Romano, Michael, 145

Rosebud Sioux Reservation, South Dakota, 22

Rosendahl, Stan, 146–47

Rowse, Doug, 84

runoff, 123–24

rural activism, 93–94

Saline County, 95

salmonella, 121

Samuelson, Drey, 59–60, 61, 62

Sand, Chuck, 50, 67, 75, 77, 88, 89, 92, 97–98, 114–15

Sand Livestock Systems: applications for permits by, 42, 78, 102, 136; carcass disposal by, 47, 107, 108; economies of scale in, 16; environmental record of, 19, 25–26, 31, 76, 77, 88, 101, 118, 119; lobbying by, 51, 96, 98; protest against, 92; sites for confined hog operations by, 13, 14, 30, 65, 75, 113, 115, 137; "SLAPP" lawsuit by, 138

Sandoz, Mari, 91

Sands, Dave, 106

Sand Systems, Inc., 65

Save Our Rural Resources (SORR), 88, 90, 94, 117

Schellpeper, Stan, 49, 50, 52, 97, 99, 127

Schimek, Diana, 106

Schmitt, Jerry, 34, 35, 36, 46–47, 52, 99

Schmitt, Loran, 97

Schooley, Carol, 79, 90

Schooley, Ron, 21, 23, 26, 35, 79, 80, 80-7

Schrock, Ed, 46, 93, 106, 109, 115

Schueth, Dennis, 122

scours, 20

Seaboard Corporation, 57, 64

Seaboard Farms, Inc., 84, 85, 89

Seacrest, Susan, 102

Seven Springs water, 86

Seward County, 23

Sierra Club, 35, 89, 138

Sitzman, Larry, 27, 49, 97, 98, 115

"SLAPP" lawsuit, 89, 138

Small Farms Cooperative, 146

smell/odors of hog farms. See air quality and confined hog farms

Smith, Virginia, 97–98

South Dakota, 38, 58, 66, 100, 138

South Dakota Supreme Court, 42

"South Divide," 126

sow pens, 5

Spalding, Roy, 120–21

Spenner, Aaron, 80-4

Spenner, Bob, 146, 80-5

Spotlight on Pork, 16

Stanton County, 23

Stanton County Livestock Feeders Association, 49

State Game and Parks Commission, 86

St. Edward NE, 25

Stenberg, Don, 35, 56, 98, 100

Stephens, Brad, 110

Stephens, Earl, 110–11, 137

Stephens, Kathleen, 110–11, 137

Stinking Water Creek, 88

Stone, Lowell, 25

Strange, Marty, 67, 81

Stuhr, Elaine, 97

Successful Farming, 4, 21

Sun Prairie, 37, 101

surface water. *See* water resources

Tarnov NE, 137
Texas, 38
Thoendel, Dennis, 83
Thoendel, Elaine, 83, 84, 87, *80-10*, 99
Thomas, Alan, 77
Thomas, Tom, 30, 32, 80
Thomas NE, 77
Thompson, Jodi, 101
Thompson, Nancy, 89
Thone, Charles, 60
Thu, Kendall, 41, 93
Title 130, 83, 107
total reduced sulfur (TRS), 130–31
Travelers Insurance, 61
Trimble, Cleve, 100
Turkey Creek, 78
Tvrdy, Edward, 62
Twin Platte Natural Resources District, 113, 118

Uhlir, Darin, 8
Union of Concerned Scientists, 141
University of Iowa, 121, 131
University of Nebraska, 12, 16, 26, 107, 120
Upper Elkhorn Natural Resources District, 122
Upper Republican Natural Resources District, 37, 38, 101
U.S. Army Corps of Engineers, 29, 116, 124
U.S. Constitution, 62
U.S. Court of Appeals, 63
U.S. Department of Agriculture (USDA), 15, 47, 103, 141, 142
U.S. Department of Justice, 91
U.S. Environmental Protection Agency (EPA), 7, 15, 23–24, 43, 46, 91, 116, 124, 138
U.S. Geological Survey, 121, 124
U.S. House and Senate Agriculture Committees, 139
Utica NE, 96

Valentine NE, 85
Veneman, Anne, 103
Verdigre Creek, 120
Von Essen, Susanna, 132–33

Wahpeton ND, 14
Waldo, Max, *80-2*, 146
Waldo, Willard, *80-2*, 95–96
"Wal-Marting of agriculture," 35, 78–82
Walthill NE, 16, 81, 82
Washington County, 137
waste management in confined hog farms: with center-pivot irrigation, 7, 39, 44, 107, 123; with lagoons, 6–7, 53, 117–23; legislature review of, 44–55; waste spills of, 15, 26, 76
Waterkeeper Alliance, 89
Water Quality Report, 123
water resources: and contamination in surface water, 15, 45–46; monitoring quality of, 7, 43, 45, 72, 77, 78, 112–24; pathogens in, 121; and pollution risks in groundwater, 45–46, 69, 112–13
Waters, Alice, 145
Wauneta Breeze, 88, 102
Webster County, 23
Wehrbein, Roger, 36, 52, 97
Weida, Bill, 133
Weinzweig, Ari, 145
Weiss, Don, Jr., 102
Wells, Rich, 26–27
Wetovick, Kevin, 9–10, 13
Whiteing, Richard, 92–93
Willets, Dan, 19, 22
Williams, Larry, 47
Willis, Paul, 142–45
Wills, Robert, 108
Winnebago Reservation, 22
Wisconsin, 58
Witek, Kate, 98
Wolbach NE, 23
Wolbach Foods, 7–8, 9, 79, 80
Women in Farm Economics (WIFE), 60
Wood, Randolph, 24, 25, 27, 45, 46, 50, 97, 119
Woollen, Terry, 33
World Health Organization, 141
Wright County (IA), 21
Wyoming, 66

Yorkshires, 9

Ziems, Donna, 83, 84, 85, 87, *80-10*, 90, 99
zoning issues, 31–43, 136–37

IN THE OUR SUSTAINABLE FUTURE SERIES

Volume 1

Ogallala: Water for a Dry Land
John Opie

Volume 2

Building Soils for Better Crops: Organic Matter Management
Fred Magdoff

Volume 3

Agricultural Research Alternatives
William Lockeretz and Molly D. Anderson

Volume 4

Crop Improvement for Sustainable Agriculture
Edited by M. Brett Callaway and Charles A. Francis

Volume 5

Future Harvest: Pesticide-Free Farming
Jim Bender

Volume 6

A Conspiracy of Optimism: Management of the National Forests since World War Two
Paul W. Hirt

Volume 7

Green Plans: Greenprint for Sustainability
Huey D. Johnson

Volume 8

Making Nature, Shaping Culture: Plant Biodiversity in Global Context
Lawrence Busch, William B. Lacy, Jeffrey Burkhardt, Douglas Hemken, Jubel Moraga-Rojel, Timothy Koponen, and José de Souza Silva

Volume 9

Economic Thresholds for Integrated Pest Management
Edited by Leon G. Higley and Larry P. Pedigo

Volume 10

Ecology and Economics of the Great Plains
Daniel S. Licht

Volume 11

Uphill against Water: The Great Dakota Water War
Peter Carrels

Volume 12

Changing the Way America Farms: Knowledge and Community in the Sustainable Agriculture Movement
Neva Hassanein

Volume 13

Ogallala: Water for a Dry Land, second edition
John Opie

Volume 14

Willard Cochrane and the American Family Farm
Richard A. Levins

Volume 15

Raising a Stink: The Struggle over Factory Hog Farms in Nebraska
Carolyn Johnsen

Volume 16

The Curse of American Agricultural Abundance: A Sustainable Solution
Willard W. Cochrane